Exploring REALITY and its Uncertainties

This book is dedicated
to the memory of our daughter
Sara Cohen-Krausz
a brilliant scientist,
an inspiration to the family
and to all who knew her.

Exploring REALITY and its Uncertainties

ERNEST KRAUSZ

sussex
ACADEMIC
PRESS
Brighton • Portland • Vancouver

2 4 6 8 10 9 7 5 3 1

First published 2010 in Great Britain by
SUSSEX ACADEMIC PRESS
PO Box 139
Eastbourne BN24 9BP

and in the United States of America by
SUSSEX ACADEMIC PRESS
920 NE 58th Ave Suite 300
Portland, Oregon 97213-3786

and in Canada by
SUSSEX ACADEMIC PRESS (CANADA)
90 Arnold Avenue, Thornhill, Ontario L4J 1B5

British Library Cataloguing in Publication Data
A CIP catalogue record for this book is available from the British Library.

Library of Congress Cataloging-in-Publication Data
Krausz, Ernest.
Exploring reality and its uncertainties / Ernest Krausz.
p. cm.
Includes bibliographical references and index.
ISBN 978-1-84519-350-8 (h/c : alk. paper)
1. Reality. 2. Knowledge, Theory of. 3. Knowledge, Sociology of. I. Title.
BD331.K656 2010
110—dc22

2009031753

FSC Mixed Sources
Product group from well-managed forests and other controlled sources
www.fsc.org Cert no. SGS-COC-2482
© 1996 Forest Stewardship Council

Typeset and designed by Sussex Academic Press, Brighton & Eastbourne.
Printed by TJ International, Padstow, Cornwall.
This book is printed on acid-free paper.

Contents

Preface and Acknowledgements

This book is about us and reality: about us as individuals, the societies we have created, the world around us and the whole universe. It is in this sense that I define reality as 'a boundless totality'. The human mind can conceive reality but cannot understand it fully. One can look at it from different vantage points and attempt to understand it with the help of philosophical ideas and scientific explanations. The possibility of using such tools tempted me to explore reality and to analyse it, at least many of its aspects, from an interdisciplinary approach. Our difficulty to understand reality fully is due to our partial knowledge of what reality truly is: the unobservables, the impenetrable aspects, the impossibility of knowing all the interconnections and causations, and certainly the inability to foretell the events yet to be caused and the elements yet to be created. One of my pivotal arguments is that human knowledge must constantly contend with uncertainties and thus individuals, societies, and mankind as a whole are often thwarted from achieving their goals. This is true in the scientific world as well as in everyday life. My argument could be countered, however, with the notion that rationality, which the human mind can apply to our lives and knowledge, could ultimately overcome the uncertainties. This argument is not convincing: in Chapter 5 I show that even in deliberate human action, in the choice of which rationality plays a major role, chance, too, is an important factor – and an unavoidable one. Thus the individual facing the world cannot really escape its uncertainties. Another argument I put forward is that, despite the advance and phenomenal accumulation of knowledge, mankind is unable to change the basic nature of reality. There are those, however, who maintain that human creativity *has* changed the human condition by producing more advanced civilizations, and that in such conditions societies emerge with higher degrees of justice and solidarity. Again, in my conclusions I express far less optimism: in my view such changes are impressive, but only in the details and not in the essence of reality.

In writing this book I realized how much I owe to the scholarly work of others. I wish to mention only a few scholars whose ideas have influenced me a great deal: Karl Popper and Isaiah Berlin, the prominent scholars of the twentieth century; and, from the 1980s onwards, I benefited immensely from the scholarly works of Nicholas Rescher and John Searle. There are, of course, those with whom I collaborated closely and who gave me much encouragement in my academic work over many years. Thus I am grateful to Julius Gould, my mentor, and to Steven Miller, my close colleague in Britain; and in more recent years in Israel to Gitta Tulea at Bar-Ilan University and to Gabriel Barhaim at the Netanya Academic College. I am tremendously grateful to Anthony Grahame, Editorial Director of Sussex Academic Press, whose guidance has been invaluable. I also wish to thank Manuela Tecusan, SAP copy-editor, for the highly professional editing of my text. And, certainly not just in a customary manner, I express my great indebtedness to my wife Gillian, who provided me constantly with wise criticism and crucial advice.

I am pleased to give special acknowledgements to the following publishers and institutions: Peter Lang Publishing, New York, for granting permission to reprint the chapter entitled 'The Cosmos – from Evidence to Speculation' from Ernest Krausz, *The Limits of Science* (2000), which appears as Chapter 3 in this volume; Wiley-Blackwell Publishing, for granting permission to reprint the article entitled 'The Elements of Rationality and Chance in the Choice of Human Action' from the *Journal for the Theory of Social Behaviour*, Vol. 34, No. 4 (2004), which appears as Chapter 5 in this volume; and Dr Henry T. Stein, Director of the Alfred Adler Institutes, for giving me permission to quote from the article by Sophia de Vries, 'Some Basic Principles of Individual Psychology', Alfred Adler Institutes of San Francisco and Northwestern Washington, 19 January 2006 (originally published in the 'Individual Psychology Bulletin', Vol. 9, 1951). I am especially thankful for the permission to use the British Library of Political and Economic Science at the London School of Economics and the Philosophy Library at Bar-Ilan University. Similarly, I wish to express my gratitude for the full support I received from Bar-Ilan University and from the Netanya Academic College in Israel.

Finally, two explanatory notes for the reader. First, given the relatively autonomous status of each chapter and the choice of grouping all page references in one place, at the end (rather than dispersing them throughout the text), it seemed appropriate to separate the bibliographic references chapter by chapter instead of merging them into one unitary bibliography. Secondly, the choice of the masculine pronoun, which occurs in all the examples of this book, together with the preference for the vocable 'man' to stand for 'mankind' and 'human being', are purely conventional and in no way reflect gender bias or discrimination. Any such attitude would be completely alien and unacceptable to the author of this book.

ERNEST KRAUSZ

1 What is Reality?

Karl Popper, the famous twentieth-century philosopher of science, held that the real world is that 'which kicks back at you'. I wish to diverge from this purely empiricist stand by focusing on certain dimensions of reality which are the result of human creation. This, of course, raises the question whether there is an intrinsic reality *ab initio* – in Immanuel Kant's conception, *das Ding an Sich* – which exists both in the physical and in the human world. To put it another way, is there an intrinsically absolute and true reality? Such a reality is difficult to define or pin down, since our understanding and knowledge both of the physical and of the human world is limited, and it is very unlikely that humanity will ever achieve a perfect and unlimited understanding of its surroundings and of the phenomena unfolding in them.

In my view, the reality we can relate to is one which is partly our creation. We create reality in different ways, and in all its aspects. The findings of science in the natural world provide us with a picture of reality which, by the very nature of scientific work and its advance, is constantly changing and it relates to only a part of the myriad connections which make up the total reality. This is also true when we consider the scientific study of the human social world. What is at play in this creativity is the individual cognitive understanding of what one actually observes, or of what one is told about phenomena, as well as the influences of the social interactive world. It can be argued therefore that, both in the physical and in the human world, that which is seen or experienced by the scientist or by the layman is a matter of his or her evaluation and interpretation, and is subject to the meanings attached to phenomena or objects which are either observed directly or else experienced in other ways than through observation. This means that the reality we can know is subject to variations between the reality seen, experienced and understood by one scientist (or by one group of scientists or laymen) as against the reality comprehended by another. As a result, we may say that 'reality is what each person *thinks* is reality' according to his or her knowledge, personality makeup, experiences, moral values, religious or secular beliefs, artistic inclinations, imaginative powers, and social behaviour, and also according to the way such-and-such characteristics of his or her parents, siblings, other members of the community and society, and wider global events and phenomena influence each person.

I have quickly proceeded above from defining reality in terms of extreme empiricism to conceiving of it in terms of extreme relativism. However, one could argue that some intrinsic reality does exist beyond what each person can see, define and interpret. Thus, does it matter that light is deemed to be composed of particles or of waves, if in any event it facilitates vision? In other words 'light is light', whatever you call it and in whatever way you understand it; and this fact is true reality. Or does it matter that we consider such-and-such vegetable as food or as vitamins, as long as its consumption, whatever we call it, provides energy for human beings and the animal world? So foods and vitamins are basically *the same* thing: intrinsic reality! Similarly, a revolution or a war could reflect class struggle, or a rise against tyranny, or a conflict between nations. Its processes and results – death and destruction, and perhaps a new order of greater equality and democracy – could be regarded as intrinsic aspects of such events in human reality.

Having introduced the most complex aspect of understanding what reality is, I shall now present a few approaches to this problem.

Isaiah Berlin's Sense of Reality

Isaiah Berlin expresses his concept of reality through a critique of the views held by thinkers during the period of Enlightenment as to how historical reality can be understood and how social and political reality could be reshaped. His basic premise is that there are some criteria of reality which enable us to distinguish 'the real from the illusory'. These criteria are produced by the methods we use in our perception and understanding of what is real, both in the case of physical objects and in that of societal structures or social events, as against what we fancy or imagine; the latter leading to escapist and utopian thoughts, which do not grasp reality as it is or as it could be. We cannot understand history, or life, or the world, Berlin says, unless we grasp 'a particular set of laws and formulae' which one or another of the many schools of thought propose. Each school offers such a set 'as the key to its explanation of why what happens must happen as it does and not in some other order. What is common to all these schools is a belief that there is an order and a key to its understanding' (Berlin, 1997). Various schools of thought gave us different vantage points: the Marxist emphasis on the economic factor; the Durkheimean stress on unintended social patterns; or the Freudian importance accorded to the irrational and to unconscious factors in the makeup of an individual. But none of these deterministic patterns provided us with a key to truly grasping the reality of the past, of the currently unfolding events or of the future shape that reality will take.

There have been great system-builders, whether metaphysical, religious or scientific, who have 'both expressed and influenced human attitudes towards the world' and thus have created 'the light in which events are seen' (ibid.). Yet, when it comes to historical events or periods, Berlin denies that it is possible to recapture such historical realities truly. Similarly, there were those thinkers, especially in the eighteenth and nineteenth centuries, who, impressed by the limits of free human action and by the regularities displayed in nature and in the functioning of human bodies and minds, conceived the possibility of an empirical science of history and envisioned the latter's inexorable march – an evolutionary trend with its own laws which, like the physical sciences, would lead to a predictable development which could not be reverted to the earlier stage. The ruthless dictators of the twentieth century proved otherwise – namely that reality could become something contrary to 'the laws of advancing civilization'. Berlin argues convincingly that the reality of past periods cannot be fully reconstructed or properly understood, nor can future realities be foretold.

This is not all. It is not just that we are not able to see the past reality exactly as it was at the time, exactly as it was seen by those steeped into the culture of their own day, and that we are even less able to imagine a future reality, which we could not conceive of because we lack the knowledge of future events – but even of our current reality we can have only a partial picture. As Berlin puts it:

> In trying to acquire knowledge about the world, external or internal, physical or mental, we inevitably notice and describe only certain characteristics of it – those which are, as it were, public, which attract attention to themselves because of some specific interest which we have in investigating them, because of our practical needs or theoretical interests. (Berlin, 1997)

Furthermore, the knowledge of the reality we discover depends not only on our own habits and beliefs, but also on that of others. We take for granted much of what we discern on the basis of our experience and past knowledge, yet we are aware only of a fraction of the facts – and of the beliefs, attitudes, angles and many other notions in which such facts are grounded – which we continually apprehend. We become aware of how little of reality we can take in when we consider the simplest act of an individual, itself based on that person's wishes, needs, circumstances and so on, and especially when we confront the complexity of the natural sciences, social sciences, history, literature, art and so on. This, Berlin says, 'is not a matter for surprise or regret: if we were aware of all that we could in principle be aware of we should swiftly be out of our minds' (ibid.). And much of what we *can* be aware of, and thus know and interpret, depends on imaginative insight. Yet this is not to say that we do not distinguish between 'facts' and the 'interpretation of facts'. Facts 'can be demonstrated by the evidence, as opposed to interpretations, theories, hypotheses, perspectives' (ibid.). Whether in history or in science, all will agree about the existence (past or present) of certain facts such as a particular war or an eclipse of the moon. There is no place here for relativism, but the facts alone do not make up history or science; the facts must be put into a special context, which is impregnated with intersubjectivity, interpretations and theories. This is necessary in order to see and understand the 'real life' situation and its causes. This is the way we face reality and cope with it.

Finally, although reality can be seen as containing certain laws and trends and uniformities, a great deal of it remains 'a dark mass of factors'; not even the sciences can grasp everything, and this is certainly true in human affairs. We are left with many unknown and half-known factors and with a great deal that we surmise; and there is much that is ungeneralizable and must be treated as coming in *ad hoc* situations. Much of reality remains incalculable and its ways are uncertain; hence Isaiah Berlin is against viewing reality in a dogmatic manner. Besides, the latter springs from misconceptions and fanciful ideas. The medium in which we live is, in this sense, a difficult one to comprehend fully (or even partially), and this often leads to a lack of realism on our part. The overall picture we gain from Berlin's own 'sense of reality' is one in which he outlines, with superb intellectual sensibility, the need to comprehend reality with utmost flexibility and common sense.

Philosophical Views

Philosophers like René Descartes (1596–1650) and Immanuel Kant (1724–1804) tried to understand how the individual connects up with the reality in which he exists, and wrestled with the question whether, as human beings, we can know what reality is in its essence. For Descartes it was a mystery how these two distinctly different domains can interact: the human mind, exhibiting consciousness and thinking, and the human body where it is located, which is part of the physical domain. Yet it is this very interaction that makes us conscious of reality. It appears that the senses are just media through which the physical objects and events are translated into mental events transmitted to us. For instance the knock I received when bumping into a physical object is translated into a feeling of pain. Furthermore, it is in this way that I am most obviously conscious about my existence. In the final analysis, however, Descartes' dualism is overtaken by his supreme dictum *cogito ergo sum* ('I think, therefore I exist'), which disperses the doubt about my very existence and thus also proves that the world I live in is real, not illusory.

But it is interesting to see that Cartesian dualism was accepted by many philosophers. Thus John Locke (1632–1704) delved into the question of how, through our senses, which are affected by objects in the world, sensations are created which in turn become impressed on our minds. It is in this way that we attain knowledge of reality. According to Locke,

> knowledge originates from two sources – sensation and reflection. Sensations are 'the impressions made on our senses by outward objects that are extrinsic to the mind'. When the mind 'comes to reflect on its own operations about the ideas got by sensation and thereby stores itself with a new set of ideas' it gets ideas of reflection. Knowledge is henceforth a process of compounding, of repeating, comparing, and uniting sensation. (Dewey, 1969)

In Locke's view, therefore, man's understanding of reality is based on this process.

A contemporary of Locke, Gottfried Wilhelm Leibniz (1646–1716), was critical of Locke's views on how knowledge of reality is attained and what the true nature of sensation is. In Leibniz's view, 'a sensation is not an effect which one substance impresses upon another'. It is not that our senses are picking up a more real world underlying our immediate experience: as a phenomenalist, Leibniz denies that there are objects which are possibly different and separate from our perceptions and mental constructs. For him, reality constitutes a system, and this is 'an organic whole – not two parts with a chasm between them' (ibid.). Departing from dualism, he maintains that everything, whether material objects, human thought, or motion, in fact the entire universe, consists of the same basic simple substances, whose nature is action towards a goal; and he calls these substances 'monads', using an ancient Greek noun (*monas*) which meant 'unit' or 'monad' and derived from an adjective meaning 'one', 'single', 'unique', sometimes even 'supreme' (*monos*). Each monad is 'thoroughly individual, having its own unique mode of activity'. However, 'its content, that which this activity manifests, is not peculiar to it as an individual, but is the substance or law of the universe. It is the very nature of the monad to be representative. Its activity consists in picturing or reproducing those relations which make up the world of reality' (ibid.). John Dewey points to Leibniz's mode of reconciling individuality and universality by quoting his own words:

> Each monad contains within itself an order corresponding to that of the universe – indeed, the monads represent the universe in an infinity of ways, all different, and all true, thus multiplying the universe as many times as is possible, approaching the divine as near as may be, and giving the world all the perfection of which it is capable. (Ibid.)

And Dewey adds: 'The monad is individual, for it represents reality in its own way, from its own point of view. It is universal, for its whole content is the order of the universe.'

One is obviously struck by Leibniz's ideas about reality in that they seem to be a precursor to the most advanced scientific theories of today concerning the structure of the universe and the particles of which it is composed: molecules, atoms, protons, electrons, and so on. Even more, his evolutionary idea that the differentiation between various aspects of the universe – or reality – represents the different levels of action and development involved seems to fit modern scientific notions of simple combinations of particles as against more complex ones, and of stages of development in the physical and psychological realms. The parallel is there between 'primary particles' which compose the visible universe and Leibniz's philosophical monads, in that neither the monads nor the primary particles can be observed adequately by our physical senses, although 'we can establish [the] existence [of the latter] by experiments which show the effects they produce' (Ryan, 1963). Nevertheless, the sceptic who ponders upon Leibniz's world – which, to Leibniz himself, is illusory yet fathomable through his theosophical notions of mysterious monads enabling one to search for a supra-sensory reality – could well ask: Do not these notions actually point to a view about the underlying nature of reality? And are they not simply a fancy of his imagination?

Let us now consider, albeit briefly, the profound ideas of another great philosopher, Immanuel Kant (1724–1804), who combined in his philosophy a number of existing theories, especially rationalist and empiricist ones. Thus Kant held that there can be no innate ideas without some prior experience. But he also held that not all knowledge always emanates simply from sense experience and is therefore always bound to objects which exist in reality. He suggested that 'the apparatus of human sensibility and understanding, that is the way in which we perceive, identify, and reflect upon objects might itself have a form or structure which in some way moulds or contributes to our experience' (Flew, 1979). He went further and claimed that some concepts were not learnt from experience, but were thought of and understood independently of experience and were applied to the latter. In other words, human beings can apply 'pure reason' or 'a-priori reason' to reality. His conclusion was that the senses do not enable us to know the things themselves but provide us with representations of things; therefore we cannot know *das Ding an Sich*, 'the thing in itself'. This could mean that part of reality is forever hidden from us.

As I said in my opening sentence to this section, the philosophers I referred to above were preoccupied with the question of how we, as human beings, are connected with our surroundings; how we understand what goes on around us; and, basically, what knowledge we have about the reality we live in. This preoccupation has continued into more modern times, and we shall come across some novel ideas and new philosophical approaches. I shall bring here just a few examples, the first being that of the great nineteenth-century American philosopher Charles Sanders Peirce.

It is most difficult to encapsulate in a few passages Peirce's wide-ranging ideas about physical and human existence and thus glean his basic conceptions regarding reality. Nevertheless, after presenting briefly some of the well-known aspects of his philosophy, I shall try to gather the strands together and thus attempt to understand his view of reality.

In Peirce's philosophy we find a distinction between three categories:

> The first [category] comprises 'the qualities of phenomena, such as red, bitter, tedious, hard, heartrending, noble'. This is the category of *feeling*, these elements being grasped by our senses. 'The second category of elements of phenomena comprises the *actual facts*'. This actuality is the external real world which 'forces its way to recognition as something *other* than the mind's creation. This is reality, or

> the world of fact, of experience. The third category is 'the medium or connecting bond' between the first and second categories. This is where thought and consciousness come into play. It is this triadic relationship that involves meanings and brings about tendencies and reactions in the future. (Buchler, 1956)

As to how the endless temporal process works in reality, Peirce proposes a semiotic theory – a three-way process of sign relationship: the sign, the object signified and our thoughts, which connect them. This basic view leads Peirce to uphold 'the old truth that man is a social animal' and to see reality in pragmatic terms, which recognizes that 'individual reaction or private sensation are incommunicable', our communication with others being based on that which is 'public and general'. 'If our language is to possess cognitive meaning, it must be defined by ways in which it is used communicatively' (Buchler, 1956). Therefore 'reality is never grasped with certainty in any special case, but only generally', and the test of certainty is to be found in the shared consciousness of a community (Moses, 1995). Peirce saw uncertainty as prevailing in the whole of reality – including scientific reality, which is full of reversals and contingencies. Despite this, he held that the special sciences, based on logic and experimentation as thery were, constituted valid instruments in our quest for knowledge of the external world. Similarly, he adopted a particular phenomenological approach for the empirical and scientific testing of the theories of metaphysics – which, he claimed, can be submitted to observation and analysis, difficult as this may be. In this sense Peirce put philosophical study on the same footing as the scientific study of reality.

The two concepts which are essential to his philosophy are pragmatism and chance. Pragmatism seems to be overstated by him, as he holds that an idea only has meaning insofar as it has a practical consequence, thus denying that a complex thought that does not lead to action has any real meaning just by itself. But even concerning ideas which have practical consequences, we can never tell directly what these consequences are in the real world – that is, we can never tell what consequences can be attributed directly to an idea, since ideas combine with other ideas, sometimes accidentally and in different circumstances, to produce some consequence or other (Carvalho, 2005). Peirce's very notion of chance and diversity should attest to this.

It is clear that Peirce's overall view of reality is pluralistic. The physical world is evolutionary, its endless process being bounded by two extremes: chance functions as a limit towards one end, rigid necessity does the same at the other end. In this evolutionary process, there is a generalizing tendency towards the regularities of the universe. As to the human world, although Peirce's philosophy is not dualistic, he did show a special interest in it and in the mental side of reality. With its many expressions, rules and habits, this world is the outcome of the shared consciousness of different communal or societal existences. In this context, too, there is a tendency towards generalization, yet diversity is ever present. In the last resort, for Peirce reality is one where the rule of mechanical necessity will always meet with spontaneity and interference, which accounts for 'the variety and diversity of the universe' (Buchler, 1956), and for the great changes and variety of patterns that we experience in our lifeworld.

Peirce's sophisticated, pluralistic and wide-ranging philosophical ideas went beyond, and adumbrated, the need to integrate the specific ways in which logical positivism on the one hand and existentialism on the other approached reality and impacted on our understanding of the world we exist in. For logical positivism and existentialism are the two major, more dogmatic philosophical schools of thought which have developed during the latter half of the nineteenth century and throughout the twentieth. The analytic–scientific approach to philosophy was strongly countered by experiential philosophy, especially in continental Europe. Where reality was at stake, European interest was clearly in the life of the individual person and it highlighted the way in which the latter could define himself by a process arising through his very existence. Although Søren Kierkegaard, Friedrich Nietzsche and Edmund Husserl, nineteenth and early twentieth-century philosophers, are regarded as the main founders of this direction in philosophy, it was the French Jean-Paul Sartre (1905–1980) who became the prime mover of existential thinking, and for a lengthy part of the twentieth century. The more immediate inspiration for Sartre's ideas came from Martin Heidegger's 1927 *Being and Time*, 'an enquiry into "the being that we ourselves are", which he termed

"Dasein", a German word for existence' (*Stanford Encyclopedia of Philosophy*, 2005).

The basic view of the existentialists is that existence precedes essence (Sartre, 1955). It is the unique experiences and personal choices of the individual human being, throughout his existence, that define his very being. This cannot be determined and cannot be understood by means of scientific concepts such as causality or by means of logical interpretations. The human being is his own interpreter and, unlike objects or substances, in order to fathom his lifeworld we need a different set of categories, governed by the norm of *authenticity*, which would enable us to grasp human existence as a particular kind of reality and thus 'fully capture what it is that makes me *myself* my "ownmost" self' (*Stanford Encyclopedia of Philosophy*, 2005). This cannot be achieved by a rationalist empirical scientific analysis or system.Not all the existentialist philosophers were of the same ilk, but the main thrust which began in the nineteenth century and then dominated the twentieth was one of atheism and nihilism, and in particular one of rebellion against the straightjacketing of the human being and his desires, undertaken by religion and science. For Husserl the phenomenological *ego* is always present in the individual, constantly pondering as he does over his doubts, denials and affirmations in his stream of consciousness. As for Nietzsche, a person lives through his life affirmation, his will to power and his self-mastery; and this culminates in the *Ubermensch*, the 'superhuman' or the individual capable of overcoming the forces working against him. The more basic notion of existentialism, however, was put forward by Martin Heidegger, who denied that the meaning of an individual's very being can be determined by the instantiation of that individual's essence or by appealing to some pre-given framework or system. Neither can facticity – to use a term coined in English after Habermas' concept of *Faktizität* – which establishes certain physical or social property of the individual, settle *the meaning of being*, since the latter includes the stance of the individual towards his own factual being and therefore transcends (that is, goes beyond) such factual existence. Therefore Sartre insists on freedom of choice and decision by the individual, since these attributes distinctively shape his life. The individual is to make his choice according to what he truly values and irrespective of existing values, fashions, or social patterns and systems. He is to be true to his real self, since he is responsible for his actions. And, if he is to be true to his real self, he must not act in 'bad faith' (ibid.). All these ideas echo Kierkegaard, who held that the reality of everyone's existence proceeds from the inwardness of the person and the truth of subjectivity. Sartre stresses in particular that the human being is alone and can only count on himself; his destiny is what he makes of himself and nothing else matters. Yet, despite proclaiming the freedom of the individual to choose his actions and to create his own self, Sartre portrays a nihilistic and pitiful picture of human existence in the universe; for he sees human beings as alienated and abandoned in a world they did not create and facing a life which leads to absurdity and to nothingness in a meaningless universe (ibid.). This is modern existentialism's expression of what reality is.

Existentialists, as Raymond Aron points out in his critique, 'take individual consciousness as the primary reality and historical totality appears incompatible with their mode of thinking'. In his view, '[a]uthenticity – the courage to come to terms with oneself, one's inheritance and one's intimates, and reciprocity – the acknowledgment of the Other, a willingness to respect him and help him realize himself – these appear to be the two cardinal virtues of *homo existentialis*'. As Aron reiterates, existentialism is *ahistorical* and it pays *no attention to totality* (Aron, 1956); in consequence, we could regard the existentialist approach as providing only a very partial picture of reality, its central theme being the fate of the individual in the world which surrounds him, and its achievement being the attempt itself of reaching an understanding of a very basic human aspect of reality.

Some Sociological Perspectives

Certain dimensions of reality will be highlighted here through some of the writings of three contemporary sociologists: Jürgen Habermas, who has developed a theory concerning the social reality of the modern world; Peter Berger, who has emphasized the social reality of religion; and Thomas Luckmann, who, together with Berger, has written seminally on the social construction of reality.

According to Habermas, the primary crisis in modern society revolves around the following facts: human beings lack freedom; individual needs are not met; and 'institutions in society are manipulating individuals'. But people do respond to this crisis by interaction, which he calls 'communicative action' (Habermas, 1984). Habermas sees the possibility of transforming societal reality by means of rational communicative interaction. In putting forward his view on the way this can lead to the evolution of societies, he emphasizes the need to do away with mythical worldviews. This means not only doing away with the confusion between nature and culture, but also highlighting the difference between *internal connections of meaning* and *external connections of objects*. These are different domains of reality. The former is embedded in linguistic communication and is characterized by the privileged access of the individual to an internal world, where he exercises his subjective *interpretations* and orientations towards action and social interaction. The latter is characterized by physical causation: it is the relation between cause and effect. Habermas sees the individual actor as someone who deals with the situations he is facing by means of communicative rational action with other human beings in the social groups where he belongs, on the basis of solidarity and of the processes of socialization into which he has been reared (ibid.). The stress here is on intersubjectivity and mutual understanding (Outhwaite, 1994).

In the social world which Habermas examines it would appear that human beings create their own reality and, quite obviously, their understanding of that reality is through their own knowledge, ideas, orientations, wishes and interests. He is, therefore, against positivism and scientism, which see reality simply in empiricist and logical terms, assuming a prestructured reality gained through observation alone (Outhwaite, 1994). In any case, as a critical theorist of the Frankfurt School, Habermas can be regarded as one who is interested in reality not only as it is, but as it could and as it should be.

As far as Habermas is concerned, neither the empiric–analytic nor the historical–hermeneutic sciences can be accepted when dealing with reality. The former describe reality according to rules, which means that 'man acts according to constant external codes', while the latter are restrictive in that they are bound by method ological rules to unifying interpretations of reality – interpretations influenced by tradition. In both cases, 'thought patterns present an objectivist illusion' and a 'distortion is created when the above mentioned sciences claim only one possible framework of knowledge'. Hence Habermas' aim is to discover reality through a critical social science which is based on reflexivity and thus is able 'to reach a more accurate understanding of reality'. As Hanan Reiner sums it up: 'This is the scientific culture preferred by Habermas and reflects the idea that reality cannot be fully described and understood by a single conceptual framework' (Reiner, 2005).

Basing themselves on the theories of classical sociologists and twentieth century philosophers, in the 1960s the sociologists Peter Berger and Thomas Luckmann have produced a treatise on the social construction of reality which is both conceptually rich and theoretically insightful. Their definition of reality is not unlike that of Popper, which has been mentioned at the beginning of this chapter: 'a quality appertaining to phenomena that we recognize as having a being independent of our own volition (we cannot "wish it away")'. The second basic point is that 'there is some sort of relationship between thought and an "underlying" reality other than thought'. And the third point which deserves to be stressed is 'the fact of *social relativity*: What is "real" to a Tibetan monk may not be "real" to an American businessman' (Berger and Luckmann, 1972; my emphasis). Put succinctly, the very fact of the multiplicity of social realities gives us a strong hint that all human societal reality is socially constructed. But how does this social construction come about?

The human being, who is born into an existing social reality, has to engage in his everyday life through interaction with other human beings and through subjective interpretations of the meanings he attaches to his own interaction or to that of fellow members of his society. His consciousness enables him to carry out these actions and to make the interpretations in different spheres of reality. Much of this takes place according to existing patterns. The existing language, knowledge, and order of things impose upon human beings their specific manner of coping with, and of understanding, the processes of everyday life. All have to go through the same processes of interaction, communication, and meaningful interpretation of these processes; and, although these do not fully overlap, there is enough common ground among human beings for them to be able to continue the processes necessary for the maintenance of the social structure thus established. The latter contains institutions, roles, rules of conduct and so on – which appear as an objective reality. Furthermore, there is also an 'objectivation of subjective processes and meanings by which the intersubjective common-sense world is constructed'; an example would be the fact that, at times, the anger felt by an individual is translated into violent acts. What is experienced in everyday life is therefore a *compelling facticity*, which is reality writ large. Yet this reality is not unchangeable; it can be changed and reconstructed by means of the same processes that human beings are constantly engaged in as they are living their lives. This leads directly to the idea that, while social reality is objective and given, it is at the same time socially defined and structured (Berger and Luckmann, 1972).

Peter Berger continues the idea of reality being socially constructed by applying it to the realm of religion. In his work on 'the social reality of religion' he emphasizes 'the fundamental dialectic process [which] consists of three moments or steps'. These are: an externalization which is seen in the ongoing outpouring of human physical and mental activity into the world; an objectivation which consists in the production, through this activity, of a reality which becomes an external fact for human beings; and an internalization of this objective world, which once again enters the subjective consciousness of human beings. 'It is through externalization that society is a human product. It is through objectivation that society becomes a reality *sui generis*. It is through internalization that man is a product of society' (Berger, 1973).

Berger holds that religion enters the world of human beings within the above framework. Thus people create an order concerning the things and phenomena they encounter, and this order becomes established and is taken for granted. This *nomos* is extended and merges into a wider universal conception, and thus *nomos* and *cosmos* become co-extensive. The 'nature of things' is then viewed cosmologically, and its stability is deemed to be derived from more powerful sources, which transcend the everyday human world. This leads to the idea of a division between the sacred and the profane. As Berger puts it: 'Religion is the human enterprise by which a sacred cosmos is established . . . The sacred cosmos is confronted by man as an immensely powerful reality other than himself. Yet this reality addresses itself to him and locates his life in an ultimately meaningful order' (Berger, 1973). At a deeper level, the sacred order also helps to defend social life against the horrible alternative of chaos and anomie, and by means of religion a legitimation is obtained for the world order which is necessary for its maintenance. But order sanctioned by religion is in fact a reality which has been defined in a particular human collectivity.

> It can thus be said that religion has played a strategic part in the human enterprise of world-building. Religion implies the farthest reach of man's self-externalization, of his fusion of reality with his own meanings. Religion implies that human order is projected into the totality of being. Put differently, religion is the audacious attempt to conceive the entire universe as being humanly significant. (Ibid.)

One could interpret Berger's ideas further. Religion may be viewed not only as human imagination and social construction. The emphasis could be placed on the deep-seated need of human consciousness to dichotomize the world into an ephemeral realm – that of everyday life – and an other-worldly, eternal and supernatural realm, the awesome power of which is beyond human comprehension. But whatever meanings are attached to reality, once they are expressed in religious beliefs these meanings become themselves part of human reality. People's belief, say, in divine revelation is an integral part of reality just as much as their willingness to take medication or undergo surgery.

In modern society religion has undergone a process of secularization, which Berger defines as 'the process by which sectors of society and culture are removed from the domination of religious institutions and symbols'. Among secular attempts to present and explain human existence and the existence of the entire cosmos, the most important is that of modern science (Berger, 1973). It is to its view of reality that we turn our attention in the next section.

Comprehending Reality through Science

I shall now consider the logical positivist view of reality, which has had a tremendous impact on how human beings understand what goes on in the world and how its processes can be explained. According to logical positivism, a reliable presentation and a true understanding of reality must be based on the systematic reduction of human knowledge about the world to logical truths and scientific foundations. Well known for this kind of approach were the philosophers of the Vienna Circle under the leadership of Moritz Schlick, who advocated 'consistent logical and mathematical reasoning, together with cautious observation of nature' (Kemerling, 2001 a). A very large number of analytic philosophers, mathematicians and scientists contributed, especially during the earlier part of the twentieth century, to this line of approach. Thus Alfred North Whitehead and Bertrand Russell developed and popularized modern mathematical logic. Bertrand Russell was influenced by Gottlob Frege's logical insights, which he combined with David Hume's empiricism; as a result, he came to regard the world as being composed of atomic facts. His ideas in turn influenced the early writings of Ludwig Wittgenstein, who shared with Russell 'the conviction that the use of mathematical logic held great promise for an understanding of the world'. And, since the world consists entirely of facts, human beings are aware of this reality 'by virtue of our mental representations or thoughts', which *picture* the way things are, and which are subsequently expressed in propositions about the nature of reality. This reality becomes meaningful when expressed in a logical language which presents the true world' (Kemerling, 2001b, paraphrasing and commenting on Wittgenstein, 1958). Others, like A. J. Ayer, Rudolf Carnap and Carl Gustav Hempel, insisted that propositions about the world must be subjected to *verifiable empirical observations* in order to be proven to be true. If such checking is not attainable, the propositions in question must be rejected. However, as Karl Popper has pointed out, an empirical verification which is final and immutable is not really possible. As we know from scientific work, scientists may at any moment come across new facts which would falsify the accepted explanation of a phenomenon. Faced with contrary evidence, the scientist is forced to search for an alternative explanation and thus to abandon the former theory. This means that the scientist constantly tests existing theories representing and explaining reality which, so long as they are not falsified, can be regarded as corroborated for the time being, but can never be fully proved (Krausz, 2000). Furthermore, as many scientists have pointed out, there are uncertainties and limitations in the ability of science to uncover fully the reality it confronts. I shall now try to portray and exemplify the way science has progressed in its endeavour to present and explain reality, despite the accompanying challenges and uncertainties it always faces.

What is special about science is that it seeks to describe and explain all the domains and all the aspects of the world by means of facts. It gathers factual evidence and it tests it through observation and experiment, nowadays using sophisticated equipment, in order to propose theories which explain phenomena and happenings in the world. Scientific work of this nature abides by certain well-known principles: objectivity, logic, and a preparedness to submit one's methods and findings for replication and validation by other scientists. As we have said above, these findings remain tentative, since new findings may come to light which may alter the explanations; yet, so long as existing theories are not refuted, what science tells us about reality may be confidently accepted. Such confidence, however, may not obtain where controversies exist regarding a given theory.

The criterion of non-controversiality is therefore crucial in science, and it is one that places the physical sciences high on the 'scientific' scale, while history and social science are much further down. Where we have observable 'facts' rather than constructions and reconstructions of assumed historical conditions, past physical states or present-day societal situations, our descriptions or theoretical explanations of reality become more solidly based. However, science as a whole (including the physical sciences) uses certain principles in order to produce its descriptions and explanations of reality – that is, its rules, laws and theories. The principles in question are the cognitive values Hilary Putnam has pointed out: coherence, simplicity and instrumental efficacy, without which science cannot work. But 'none of them is reducible to physical notions'. This means that the fact/value dichotomy breaks down in the presentation of reality offered by science (Putnam, 1990).

In its constant search for and probing of what reality consists of, science has extended the realm of observability tremendously, and has thus established, in great depth and accuracy, scientific facts about the tiniest components of reality. The bacterium and its rotating flagellum can be seen under a light-microscope, and so can the cell membrane under an electro-microscope. These are scientific facts. The coherent system of observations which result in models of the functioning of biological molecules, however, are not scientific *facts* that can be observed in their entirety, since such models include postulated, but not seen, aspects and processes. Even in such experimentally advanced areas we cannot be sure that science can provide us with a one-hundred-percent 'correct' representation of reality.

Let us turn to cosmology, where one can see how stupendously our observations of the universe have expanded. To quote Gerard Piel:

> Astronomical instruments carry perception a million times farther out into the universe . . . The tools of experiment and observation reach 10 billion times deeper into the substance the universe is made of . . . scientists have discovered entirely new orders of physical reality. Human consciousness must strain and stretch to comprehend the universe . . . (Piel, 2001)

It is perhaps this fantastic advance in particle physics and cosmology that brings science to its limits in terms of close inspection and to a distancing from the possibility of explanation and verification. Let us take the Big Bang theory, which, as cosmologists claim, explains *the genesis* of the universe. This theory propounds the explosion of a highly condensed primeval atom which took place some 15 billion years ago, giving rise to a 'dense, hot, otherwise featureless entity', from which the entire universe, as we know it today, developed (ibid.). In spite of the evidence we have today that such a cosmic event indeed took place at time zero – evidence consisting of traces of the explosion in the form of weak electromagnetic radiation coming from all over space – science cannot answer the basic question of how such an event was possible, apparently *ex nihilo* (Aviezer, 1990). For, obviously, science cannot reconstruct reality as it was at time zero!

Again, in cosmology, scientists are now telling us that research of the outer edges of galaxies suggests that gravity as we know it and the atomic content of the universe apply to only some 4 per cent of it, and that our knowledge of 96 per cent of the universe is missing, the larger percentage being possibly explained by new theories. The problem which research has discovered is that, contrary to the universal law of gravity, stars at the edges of spiral galaxies are found to be racing just as fast as those nearer the centre. Is this a sign that there may be variable gravity in the universe? No, the scientists say. Many of them prefer the theory that there is 'something else', which provides extra gravity. This special 'something' they call dark matter and dark energy, but the identity of such entities remains completely unknown. Some scientists reject these ideas: they hold that you cannot base such vital theories on so little physical evidence. As for computer simulations, which seem to support the new theories, one could say that they provide some evidence through 'virtual reality' rather than through 'real' reality (BBC Science and Nature – Horizon, 2006).

It is quite obvious that a perfect science, exhibiting completeness in that it knows everything about reality and provides a final statement on it, is unrealizable – as Nicholas Rescher (1984) has pointed out. It seems therefore that we have to accept scientific realism – in other words, 'the doctrine that science describes the real world' and that 'the orientation of science is factual and objective'. We can accept such a doctrine, but then we must also accept that it leads only to a 'best estimate' of what reality is like. As Rescher puts it:

> At the level of generality and precision at issue in the themes of natural science, we are not now – or ever! – entitled to lay claim to the scientific truth as such but only to the scientific truth as we and our contemporaries see it . . . Reality outruns our cognitive reach; there is more to the world than lies – now or ever – within our ken. (Ibid.)

Pluralistic Reality: A Conclusion

The review I have undertaken in the foregoing pages points

undoubtedly in the direction of a multiplicity of realities, which can be conceived of and whose meanings can be understood. The different angles from which reality is viewed and the different ways of understanding it lead to different views – to take one example, the historical and sociological frameworks, which look at more limited spheres of reality, as against the philosophical and scientific approaches, which spread out widely in order to apprehend a total and holistic reality. Reality – that is, various views of it – divides up further. Within each framework we find different claims as to the real essence of the world and the real meaning of our being in it. At a sophisticated, scholarly level, there appeared throughout the ages philosophers, scholars and scientists who provided a variety of often conflicting accounts of what reality is and how it can be understood best. The philosophical views themselves – some of which I described briefly, as examples of how philosophers try to understand reality – are testimony to the use of pure imagination in conceiving reality. In this sense they provide the most telling examples of creating different versions of reality, each one of which is justified on purely philosophical grounds yet lacks any empirical justification for claiming that it reaches *the* true reality.

Looking at what 'reality' means for the human race in everyday life, it is clear that, alongside the unifying factors which bind individuals to a human social reality, there are the 'individualized' realities of each one of us, stemming from our personal abilities, our powers of understanding, our life chances, and in particular our interpretations and imaginations. Individuals continue to carry with them, throughout their lives, their own created realities – which are, of course, prone to changes. This subjective individualization of one's 'realities' is basically a result of changing circumstances in the encounters between human beings and the reality (or realities) which surround them – and, consequently, the images they create for themselves. As Armin Bardel sums it up: 'Somewhere between the perceiving subject and reality there is a long and winding process of representation, along which an image is created of this reality. A new reality is created' (Bardel, 1997).

The creation of new forms of reality is a constant process: a new substance or a new theory created by a scientist becomes part of overall reality – and so do the new institutions and rules established in any human society. These permanent and variegated interactions, changing orders and moving patterns are the stuff that 'real life' is made of. Furthermore, the complex interconnections between aspects of our lifeworld and processes going on in the universe should convince us of the infinite variety of existing realities. Looking microcosmically at our finite lives, William James clearly formulated his view of life's pluralistic nature: 'Briefly it is this, that nothing real is absolutely simple, that every smallest bit of experience is a *multum in parvo* plurally related, that each relation is one aspect, character or function, way of its being taken, or way of its taking something else'. And each bit of reality could be actively engaged in any such relation, without being thereby automatically or necessarily related to other bits or aspects of reality (James, 1909a).

The relationships between the great variety of realities, and between the different aspects of these realities, can be conceived of as processes which constantly create, recreate, and maintain the encompassing total reality, or parts of it. These processes are prone to change, are subject to chance, and may result in fragmentation; yet this does not imply that regularities are not a feature of any of the parts of reality, or of the totality. This totality, which is boundless, is there absolutely and objectively. We can certainly conceive of the existence of a total reality as against nothingness. The latter makes no sense except as an antonym to total reality. Yet our grasping both of total reality and of parts of it is limited. This is due to our partial knowledge of what reality truly is: the unobservables, the uninterpretable aspects, the impossibility of knowing all the interconnections and causations, and certainly the inability of foretelling the events yet to be caused and the elements yet to be created produce overwhelming limitations. Humanity is engaged in a never-ending pursuit of discovering more and more of reality and of understanding it better. Despite limitations, humanity's achievements are great; yet we shall never know what total reality is. This is way beyond our comprehension.

Bibliography

Aron, Raymond, 2005. 'Le Fanatisme, la prudence et la foi', *Preuves*, 109 (translated by Imogen Forster as 'Fanaticism, Prudence and Faith'; also

in *Commentaire*, No.109, Paris, printemps 2005, pp. 5–19),
Aviezer, Nathan, 1990. *In the Beginning*, Hoboken, NJ: Ktav Publishing House, p. 9.
Bardel, Armin, 1997. 'Images of Reality: Social and Aesthetic Aspects of Representation', in Gabor Forrai, ed., *Images and Reality. Proceedings of the 1996 Miskolc Conference of the Hungarian Association for Cognitive Science*, Visegrad, January 1997, p. 5.
BBC Science and Nature – Horizon, 2006. 'Most of Our Universe is Missing', accessed 28.02.2006 at http://www.bbc.co.uk/sn/tvradio/programmes/horizon/missing.shtml, pp. 1–3.
Berger, Peter L., 1973. *The Social Reality of Religion*, Harmondsworth: Penguin University Books, pp. 14, 34–5, 36–7, 113.
Berger, Peter L., and Thomas Luckmann, 1972. *The Social Construction of Reality*, Harmondsworth: Peguin University Books (original publication 1967), pp. 13–18, 33–7, 70–89, 134, 149, 183.
Berlin, Isaiah, 1981. *Against the Current*, with an Introduction by Roger Hauseheer, Oxford: Oxford University Press, pp. l–li.
Berlin, Isaiah, 1997. *The Sense of Reality*, London: Pimlico (original publication Chatto and Windus 1996), pp. 5, 7, 14, 15, 26.
Berlin, Isaiah, 1998. 'The First and the Last', *New York Review of Books*, Vol. 45, No. 8, pp. 52–60.
Buchler, Justus, ed., 1956. *The Philosophy of Peirce – Selected Writings*, London: Routledge and Kegan Paul, pp. xi, xii, 335.
Carvalho, Olavo de, 2005. 'A note on Charles S. Peirce', accessed 25.05.2005 at http://www.olavodecarvalho.org/traducoes/ peirceng.htm.
Dewey, John, 1969. *The Early Works and Leibniz's New Essays Concerning the Human Understanding*, London and Amsterdam: Southern Illinois University Press/Feffer and Simon Inc., pp. 293, 313–15, 317–21.
Habermas, Jürgen, 1984. *The Theory of Communicative Action*, Vol. 1: *Reason and the Rationalization of Society*, translated by Thomas McCarthy, Boston: Beacon Press, pp. 48–50, 135.
Heidegger, Martin, 1962. *Being and Time*, translated by John Macquarrie and Edward Robinson, Oxford: Blackwell.
Husserl, Edmund, 1989. *Ideas Pertaining to a Pure Phenomenology and to a Phenomenological Philosophy*, translated by R. Rojcewicz and A. Schuwer, Dordrecht: Kluwer.
James, William, 1909a. *A Pluralistic Universe*, London: Longmans Green.
James, William, 1909b. *Pragmatism*, Cambridge, MA, Harvard University Press.
Kant, Immanuel, 1998. *Critique of Pure Reason*, translated and edited by

Paul Guyer and Alan W. Wood, Cambridge: Cambridge University Press.
Kemerling, Garth, 2001a. 'Logical Positivism', accessed 18.12.2005 at http://www/philosophy pages.com/hy/6qhtm, pp.1–3.
Kemerling, Garth, 2001b. 'Ludwig Wittgenstein: Analysis of Language', accessed 15.12.2005 at http://www.philosophy pages.com/hy/6s/htm, pp. 1–4.
Krausz, Ernest, 2000. *The Limits of Science*, New York: Peter Lang, pp. 8–9.
MacDonald Ross, George, *Leibniz*, 2000. Oxford University Press, Electronic Edition, University of Leeds, Electronic Text Centre (July).
Moses, Greg, 1995. 'American Pluralisms: Pragmatism, Locke and Cruse', accessed 25.05.05 at http://pages.prodigy.net/gmoses/moweb/plurals.htm.
Nietzsche, Friedrich, 2000. *Basic Writings of Nietzsche*, translated by Walter Kaufmann, New York: Modern Library.
Outhwaite, William, 1994. *Habermas – A Critical Introduction*, Stanford, CA: Stanford University Press, pp. 23, 39.
Peirce, Charles Saunders, 1997. *Pragmatism as a Principle and Method of Right Thinking*, Albany: State University of New York Press.
Piel, Gerard, 2001. *The Age of Science: What Scientists Learned in the Twentieth Century*, New York: Basic Books, pp. 2, 234.
Putnam, Hilary, 1990. *Realism with a Human Face*, Cambridge, MA: Harvard University Press, pp. 138–41.
Putnam, Hilary, 2002. *The Collapse of the Fact/Value Dichotomy and Other Essays*, Cambridge, MA: Harvard University Press.
Reiner, Hanan, 2005. *The Web of Religion and Science*, Piscataway, NJ: Gorgias Press, pp. 38–9.
Rescher, Nicholas, 1984. *The Limits of Science*, Berkeley: University of California Press, pp. 133–59.
Rescher, Nicholas, 1987. *Scientific Realism*, Dordrecht: Reidel.
Russell, Bertrand, 1948. *Human Knowledge, its Scope and Limits*, London: Allen and Unwin.
Russell, Bertrand, 2001. *The Scientific Outlook*, London: Routledge.
Ryan, Charles, J., 1963 (October). 'On the monads of Leibniz', *Sunrise Magazine* (Theosophical University Press), accessed 07.06.2005, p. 1.
Sartre, Jean-Paul, 1955. *Existentialism and Humanism*, London: Methuen, p. 26.
Sartre, Jean-Paul, 1966. *Being and Nothingness* [*L'Etre et le néant*], translated by Hazel E. Barnes, New York: Citadel Press, p. 547.
Shieber, Joseph, 1999. 'On the tenability of non-factualism with regard to the a priori', *Pacific Philosophical Quarterly*, 80, pp. 379–90.

Speake, Jennifer and Sarah Mitchell, eds, and Antony Flew, 1979. *A Dictionary of Philosophy*, London: Macmillan, p. 176.

The Stanford Encyclopedia of Philosophy, 2005. 'Existentialism', accessed 12.05.2005 at http://.plato.stanford.edu/entries, p. 2.

Wittgenstein, Ludwig, 1958. *Tractatus Logico-Philosophicus*, London: Routledge and Kegan Paul.

2 Human Knowledge and Uncertainty

It is not my intention to delve into the achievements of humanity in the realm of knowledge. Nor will I try to analyse the essence of human knowledge and its intricate byways. For generations, philosophers and scientists have considered it to be 'the be all and end all' of our very existence as human beings. For our existence, as such, is based on human knowledge. We use it to survey reality and to survive in it. We create methods which enable us to live our lives, and we possess faculties which facilitate our understanding of reality and of our very being, albeit in a limited manner. There is by now a great deal of information, produced by physiologists and psychologists, as to how knowledge itself is produced and how it works: we understand, for example, how the brain functions and what the place of consciousness is in nature (Searle, 1992). Yet the scientific limitations in this domain are still seriously thwarting the efforts. But neither such efforts nor philosophical theories about human knowledge in general will arrest our attention here. Rather, we are going to focus on the way in which human knowledge is affected by, and contends with, one specific constant in the reality we live in – uncertainty.

Human knowledge, broadly defined, consists of all that is based on, and results from, the processes going on in the human mind, the principal component of which is thinking. Such processes include awareness of the surrounding reality; intentionality and the planning of one's actions; the calculation of the effects of one's actions, which takes into account fear and repressed fear; the application of beliefs and convictions to one's behaviour; apprehensions regarding future outcomes; the recollection of past experiences, which extends to dreams, hopes for the future, and so on. Thinking itself relates not only to real, 'objective' things and events, but also to imagined ones. The combinations between all these facets of one's mental activity, the possible interruptions of the various processes, the doubts one has about one's assumptions, and the many imagined possible outcomes, or predictions of outcomes, which are linked to different interpretations of events and lead sometimes to confident decisions, and at other times to confused states – all these are instances of what may result from the process of thinking.

When one considers facets or aspects of thinking such as the ones reviewed above, and in conjunction with the fact that it is through thinking that the accumulation of knowledge comes about, it should be no surprise that knowledge is affected by uncertainty and that, more generally, uncertainties prevail in all the different types and dimensions of human experience and creativity. The reasons why knowledge is always contending with uncertainties is that human beings *do* have the faculty of choice, and that reality *is* impregnated with chance. In Chapter 5 I am going to discuss these factors, which exist in conjunction with rationality and the objective basis of reality and of human life.

I now wish to review the effects of uncertainty both in the domain of the scientific enterprise and in everyday life.

Science and Uncertainty

Inevitably, science has an anthropocentric basis and, by the same token, an anthropocentric uncertainty. For it is direct human observation, to start with, that establishes the facts; and it is the latter – the facts – that prompt the extension of these observations and thus the formation of theories. But, in order to prove the veracity of its theories, science turns to increasingly sophisticated tools of observation and measurement, and carries out repeated tests designed to induce confidence in its results and explanations.

Undoubtedly scientific instruments have allowed us 'to transcend the limitations of our native senses' (Humphreys, 2000), and the 'accuracy' achieved – that is, the close approximation of data to the true value of their source – as well as the 'precision' attained in most scientific work – that is, the fact of obtaining exactly the same result through repeated measurements – show clearly that only little uncertainty exists in this area of scientific endeavour. There are many examples on the side of achievements. Thus DNA testing – the examination of the substance which carries the organism's genetic information and through which a specific genetic profile is established in every individual –is highly reliable for any specimen we wish to identify. Fingerprinting, for instance, is less reliable by comparison, since fingerprint patterns can be altered through plastic surgery. Another good example of reliability is the case of temperature measurement. It is true, of course, that we cannot 'get up close and inspect the temperature' (Humphreys, 2000) – although we can certainly 'feel it', which is a good empirical criterion; yet the feeling of temperature is a relative matter, differing widely from individual to individual, and even from state to state. Hence we have the far superior instrument of the thermometer which, when properly calibrated, provides us with an objective and reliable grading of the temperature. The microscope is another scientific instrument which allows us to observe tiny elements in nature, not accessible to the naked eye; and the advanced electron microscope offers also the benefits of an extremely high resolution. With some methods, for instance polygraph testing for the detection of people not telling the truth, scientists are sceptical about their validity; in the specific case of the polygraph there is a great deal of uncertainty. However, with the fully accepted scientific devices uncertainty creeps in almost entirely as a result of blunders, and especially at the stage of the scientist's interpretation of the results obtained.

But in most scientific work the aim is not simply to establish certain givens of nature as accurately as possible. Let us consider a couple of examples. In the sphere of economics, accuracy in the measurement of indicators such as the consumer price index is a boon, both for the consumers and for the economy as a whole. Similarly, the images of biological macro-molecules such as protein provide accurate information regarding the density of the molecules in the samples obtained, and thus scientific research in biology is advanced. It is obvious, however, that the great value of such measurements is that they could lead to causal explanations by linking the data thus obtained to other data. Yet it is precisely in this search for causality that uncertainty enters. Firstly, there may be some weakness in the data which diminishes their reliability; in the case of the price index, not all the existing prices in the economy are included, although it is usually claimed that the index represents very reliably the overall situation concerning the general level of prices in the economy. Again, in the biological example, should a small percentage of the molecules in the sample not exhibit the findings obtained in the rest of the sample, this would be regarded as mere 'noise', and the overall findings would still be accepted as reliable. In all such cases some uncertainty remains, since in scientific research measurements of the events or objects under observation are hardly ever exhaustive: measurement cannot be conducted on all the existing members of a class. Therefore the scientist relies on randomization and sampling. The data thus obtained are usually presented in a form that shows co-variation of events, which suggests the likelihood of a causal link between the phenomena measured. This is where the task becomes most daunting. How do scientists approach this task? As Henry Pollack says, they 'operate with simplified concepts of the way things work', which we call 'models'. The latter come in the forms of conceptual, physical or numerical representations of the different aspects of the real world (Pollack, 2003). The models are then tested through experimentation.

> There is a continuous interaction between models and observation, with each undergoing adjustments in the face of the other. New observations lead to revision of a concept, and a new concept, in turn, suggests new experiments or observations to be made that will again put the concept to a test.

Pollack stresses that this iterative interplay can improve our understanding of the real world as well as reduce uncertainty (ibid.).

However, conceptual models may be built on incomplete conceptualizations as a result of lack of a full understanding, especially of complex systems, and in consequence uncertainty continues to plague scientific theories. We have witnessed this throughout the history of science (ibid.). As we shall see below, different scientists have disagreed on some of the most basic issues on account of different conceptualizations and a different modeling of the real world. This highlights scientific uncertainty. Specific problems arise in the case of physical models. For instance, wind tunnels 'utilized to study turbulence generated by different aircraft designs' do not reproduce conditions in the real atmosphere – quite apart from the fact that the scale under which experiments are carried out when they are based on physical models is different from that of the real world (ibid.).

Apart from physical models, another possibility is to use numerical models, which are 'a set of mathematical equations that describe the functioning of a system and that can generate quantitative predictions about its behavior'. A simple example would be the calculation of the monthly repayments of a bank loan, which is based on the sum borrowed at a certain interest rate for a definite period of time (ibid.). Here the conditions are clear, hence no uncertainty can occur. A numerical model applied, however, to more complex situations such as demographic trends would have to include not only objective economic and political conditions, but also subjective factors such as attitudes to family size among young people of child-bearing age, as well as unknown factors such as possible developments in the medical field which might affect older people. The uncertainties in such more complex real-world situations will still allow for the application of probabilistic models, although these will provide uncertain predictions. In sum, since scientific 'models [and experiments] cover only a restricted aspect of the real world' (Kruse, Schwecke and Heinsohn, 1991), the uncertainty of their causal explanations looms larger the greater the number of variables; and, where knowledge of real-world situations is only partial due to the existence of 'hidden variables', the model may be totally inadequate.

So far I have made reference to 'down-to-earth' problems of science which exhibit the prevalence of uncertainty. I now turn to some of the more fundamental aspects of science where, similarly, we encounter uncertainty. At the beginning of the twentieth century, the famous scientist Albert Einstein propounded his great discovery of 'relativity'. As David Peat puts it, in his special theory of relativity, Einstein

> dealt with the issue of how phenomena appear different to observers moving at different speeds. He also showed that there is no absolute frame of reference in the universe against which all speeds can be measured. All one can talk about is the speed of one observer when measured relative to another. Hence the term 'relativity'. (Peat, 2002)

Despite this revolutionary idea, Einstein was a determinist and held that there was 'a totally objective reality to the world' which was underlying the different appearances; he believed, namely, that 'the laws of nature underlie relative appearances'. Thus 'phenomena . . . are manifestations of these underlying laws but only under particular circumstances and contexts'. For Einstein, 'underlying reality is quite independent of any particular observer, [and] he felt that if the cosmos did not work in such a way it would simply not make any sense' (ibid.). It is on this point that Einstein and that other giant of modern science, Niels Bohr, parted company.

Niels Bohr's non-deterministic quantum theory was based on Max Planck's discovery that 'light energy is emitted only in discrete amounts called quanta. Rather than continuous radiation emerging from a hot body, there is a discontinuous, and finite, emission of a series of quanta'. Based on this development in physics, Bohr suggested 'that not only light, but also the energy of atoms, is quantized'. To these new ideas was added Marie Curie's discovery of radium, a radioactive element – 'which means that its nuclei are unstable and spontaneously break apart or 'decay' . . . But an individual atom's moment of decay is pure chance – it could decay in a day, or still be around after 10,000 years' (ibid.). David Peat explains and summarizes the nature of the quantum world as follows:

> Quantum chance is not a measure of ignorance but an inherent property. No amount of additional knowledge will ever allow science to predict the instant a particular atom decays because nothing is 'causing' this decay, at least in the familiar sense of something being pushed, pulled, attracted or repelled. Chance in quantum theory is absolute and irreducible. Knowing more about the atom will never eliminate this element. Chance lies at the heart of the quantum universe. (Ibid.)

While Bohr accepted such a universe, Einstein could not. There was an even greater unbridgeable gap between Einstein and Wolfgang Pauli, who interpreted the quantum nature of the physical world as 'the irrational in matter'. For Pauli, quantum chance meant that 'events . . . occur outside the limits of causality and rational physical law'. Thus a basic uncertainty entered the world of physical science. Was this reality consistent, and did it reflect at every level an objectively observable world, independent of the act of observation – or was it ridden with chance and deep down not given to explanation by rational physical law? The balance was tilted in the direction of basic uncertainty in the micro-physical realm. Werner Heisenberg explained his 'principle of uncertainty' by pointing out that an experimenter who tried to establish the position and the speed of an electron could never achieve this aim with utmost accuracy, since 'as soon as you try to measure position you change the electron's speed, and as soon as you try to measure speed you change the electron's position'. Hence 'there is always an irreducible element of uncertainty involving speed and position' (ibid.). In other words, because of the interplay between our observational instruments and the particles observed, the knowledge we acquire does not give us absolute certainty about the accuracy of our observations. Consequently, scientists have become aware of the limitations of modern science and cannot uphold a view of it which claims objectivity and certainty as its basic tenets (Heisenberg, 1958; Bohr, 1958).

I brought the above examples, as well as those in Chapter 1, with the aim to show the different perspectives and paradigms used in science in order to uncover true reality. Of course the coverage is deliberately partial and represents a sampling of the conceptual space rather than a formal map.

Uncertainty and Everyday Life

It would be stating the obvious to say that human beings constantly face uncertainty, from the very moment of conception right up to death. Even after their death there is uncertainty as to how people will remember them. Throughout their lives they face the need to make choices, but are uncertain about the outcome of their choices. In his book *Behaviour in Uncertainty*, John Cohen stresses that the main problem one encounters in choosing an opinion or an action is that of assessing the evidence (Cohen, 1964). The circumstances in which choices have to be made will greatly affect the decision-making process, and (as I argue in Chapter 5) decisions are made not only on the basis of rationality, but are also dependent on chance.

I shall try to exemplify the uncertainties of everyday life by looking at a number of areas of people's lives. The first will be 'the man–vehicle relationship in a traffic system' – which leads, in Cohen's words, to a situation of 'gambling with life on the road'. In his analysis, Cohen points to the causes of uncertainties involving the driver, his passengers and pedestrians. In the first place, the capacity of man for self-regulation is greatly reduced in the man–vehicle relationship, since even a person's normal homeostasis may be broken at the enormous speed of his mechanized travel means. Secondly, the physiological condition of stability and coordination is often abnormally affected, a fact which disturbs one's normal functioning and psychological state of equilibrium. Here emotions, often of the aggressive kind, take over, especially in relation to other drivers. Thirdly, factors which interfere directly with the driver's functioning capacity – for instance lack of sleep, exceptional fatigue, high-volume music, or electronic communication – may all contribute to reducing his power of concentration on the task at hand, namely transiting safely from place to place. Finally, traffic conditions may be unregulated or in the process of becoming so – for instance due to bad visibility in night driving, or to difficult weather conditions; and these factors add greatly to the uncertainties of modern travel. Let alone a host of other variables which can affect the driver's ability to correctly control the vehicle – such as age, level of intelligence, skill, mental state at the time of driving, the possibility of having to drive in a situation of crisis, the quality and condition of the road, and so on.

Marc Green and John Senders deal with the uncertainties that cause road accidents (Green and Senders, 2004). They look at the major causal factor, which can be summed up as 'human error'. They quote a comprehensive study, which found out that human error was the sole cause in 57 per cent of all accidents and was a contributing factor in over 90 per cent of them, as against 2.4 per cent representing accidents solely due to mechanical fault and 4.7 per cent representing accidents entirely caused by environmental factors (Treat et al., 1977). But in their view the interaction between the human factor and the situational ones is what is truly causing the accidents. It is in this interaction that a great deal of uncertainty enters. The person driving is exposed to a constant flow of information. The visual input is tremendous for such a person: the road itself, other vehicles, pedestrians, signs, satellite guiding maps, the passing scenery. In addition, there is auditory input such as listening to the radio, talking on a cell phone, making conversation with other passengers; and internal input such as remembering addresses and directions, planning the route and other chores to be attended.

> But humans have limited information processing abilities and must rely on three fallible mental functions: perception, attention and memory. When a driver fails to avoid an accident because the situation exceeds these limitations, it is often called 'human error'. In reality it is often the situation that is primarily responsible, not the driver's response to it. (Green and Senders, 2004)

It is surprising that Green and Senders reach the above conclusion. For even in circumstances of 'overload' the accident is really due to the interaction between the situation and the human operator, and not either in isolation. This makes sense because, within limits, it is open to the human operator to maximize his processing capacity by focusing attention on critical stimuli when the situation is very demanding. In any event no matter how the various factors interact in the causation of accidents, uncertainty is an inbuilt element, since the information processing ability of human beings varies so much, and the information coming their way is so varied, that assessing when and where an accident may occur is hardly possible. It is quite another matter to assess *post factum* what factors can be blamed for the occurrence of the accident.

Another area replete with uncertainties, and where human lives are at stake, is that of the diagnosis of a disease. Medical science has made tremendous progress and its achievements are in many ways phenomenal, both in the treatment of disease and in the ability to diagnose disease – or, in certain areas, even to prevent negative developments which might lead to disease. Nevertheless, a great deal of uncertainty still prevails in the area of diagnosis. A well-known case is that of Alzheimer's disease. Thus, in a study of the pathological confirmation of the diagnosis of this disease, it was shown that '[t]he positive predictive value of a diagnosis of Alzheimer's disease was 81% including coexistent diseases, falling to 44% when limited to pure cases'. The term 'positive predictive value' refers to 'the proportion of people initially diagnosed as suffering from Alzheimer's who are subsequently shown to have the disease on the basis of thorough testing'. In their conclusion, the authors point to the uncertainty that exists both in diagnosis and in prognosis concerning this disease (Bowler et al., 1998). In another study, ten other conditions are explored which may be mistaken for Alzheimer's disease (Barrett, 2005).

Turning to the topic of prenatal genetic testing, there has been a widespread demand from couples to be tested for the possibility of giving birth to a child with a particular genetic condition. Such testing provides prospective parents with an assessment of the risk of their future children suffering from certain diseases and has become the practice particularly where assisted reproductive technologies are used. I wish to quote from a presentation on this subject which points to the many uncertainties involved:

1 'Current tests can identify only a limited number of disorders . . . Even a positive test cannot always predict exactly how, with what severity or when the condition will show up. And no test can factor in the influence of the environment, medical treatment, or social circumstances and support.'
2 'Prenatal tests almost always indicate only potential, not certainty. They may say nothing about how a genetic variation will manifest itself . . . Many prenatal tests . . . leave great ambiguity as to how severely the detected condition will affect the child.'
3 'Testing can often pick out potentially serious disease very accurately, but these laboratory techniques are not foolproof' (Koliopoulos, 2001).

One may note here that the technology of existing tests is being continuously improved and new tests are developed, the aim being to improve diagnostic accuracy. Furthermore, there are guidelines for the improvement of the quality of reporting on diagnostic studies (Reitsma, 2007). Despite such efforts, it is unlikely that inaccurate diagnoses can be completely eliminated. Therefore people face the possibility of being subjected to wrong medical decisions even in cases where they detected some serious physical irregularity and have taken advice from their physician. At the other end of the scale, we are clearly exposed to the uncertainty factor where less serious conditions are at stake – conditions which appear fairly frequently, like those disseminated by germs or viruses. My catching a cold depends on a whole host of causal agents. Have I traveled in the vicinity of people suffering from a cold? Have I washed my hands sufficiently after using public facilities? Does any member of my family have a cold? Has my defence system against contagion been at a low ebb – say, due to overwork? Is my immunity level, on the whole, rather low? – and so on.

The world of uncertainty is revealed to us in all its might in the field of economic life. Early on in the twentieth century, the famous economist Frank Knight has distinguished between a 'prior probability', where all possible 'states' are defined and known – as when one rolls a number on a die – and a 'statistical probability', where the association between predicates is empirically established and one can classify instances empirically too. Where there is no basis for classifying instances, Knight held that only 'estimates' can be made. Knight saw 'the role of profit as a reward to the entrepreneur for bearing inevitable uncertainty'. The entrepreneur's estimates may be right or wrong, but, come what may, he must face up to the brute facts of economic life. Hence '[p]rofit arises out of the inherent, absolute unpredictability of things . . . a probability calculation in regard to them is impossible and meaningless'. Such absolute uncertainty is inherent in economic life. In Knight's words: 'The economic system is moving through calendar time from an irrevocable past to an uncertain and statistically unpredictable future. Past and present market data do not necessarily provide correct signals regarding future outcomes' (Knight, 2002; see also Davidson, 1994 and Janeway, 2006).

Now, speculative hedging and insurance schemes can possibly avoid or reduce some of the risks in business and financial deals;

but an element of uncertainty always remains. The basic idea that investment in one market where price fluctuations could cause losses may be offset by the simultaneous investment in another, more solid, market, where there is much more of a likelihood of gains to be made, is not sufficient to overwrite the risky nature of hedging. Turning to economic life of the kind that affects the general public, some writers have argued that an overall economic instability has overtaken even some of the advanced modern countries, for instance the United States. Jacob Hacker points to certain facts concerning the worsening condition in which retirement plans and health insurance schemes find themselves, and the growing inability of the government and great corporations to provide security in case of economic disasters. Both the family institution and the work place have been weakened as fallbacks in difficult economic situations (Hacker, 2006). Crossing a huge hiatus, we encounter the underdeveloped world where populations of hundreds of millions are subjected to the vagaries of the weather and to climatic changes, and live constantly in fear of inadequate food supplies and of water shortages. The relentless spread of the desert and drought means that many millions of people, especially in Africa's dry zone, are constantly on the threshold of death due to undernourishment. Other factors, too – for instance a very high population growth and the spread of diseases in areas such as the tropics, which are especially prone to malaria, and more generally the spread of HIV/AIDS – contribute to the fact that life expectancy in those areas is the lowest in the world. The precariousness of life is at its utmost there.

Perhaps the most poignant aspects of life's uncertainty are not those deriving from the natural conditions in which the human species exists on our planet, but those inhumanly created by humanity itself, through wars and genocide. Just to mention the number of 72 million soldiers and civilians killed in the Second World War, which was the deadliest war in human history, does not by itself evoke the total uncertainty and fear in which hundreds of millions of individuals lived during that short period of five years. The civilian losses through bombing, war-related famine and disease could affect, at any time, anyone who lived in the vast areas of the war-zones. Similarly, the total uncertainty experienced by the combatants is 'explained away', so to speak, by the very nature of combat in the battlefield, but it is most difficult to grasp when looked at from the individual combatant's perspective. The angle from which a soldier shoots and the number of bullets he releases towards soldiers of the opposing army, together with the particular position taken up by any of these combatants at any one moment, are factors which determine who is and who is not killed in battle. The route taken by a tank in a minefield, or the particular manoeuvre made by a pilot in his aircraft, can result in the death of one soldier or in the survival of another.

If we turn to genocide – a man-made situation of total insecurity, targeted at specific groups of people on account of their origin – then Holocaust survivors provide the most horrific accounts of the barbaric German Nazi genocidal machine, which brutally exterminated six million Jews and some millions of people of other ethnic groups. This colossal crime, so carefully and obsessively documented and carried out by the Nazis together with their collaborators, could be said to exemplify a very high degree of certainty, in that their plan sealed the fate of the vast majority of European Jewry. But numerous eyewitness accounts from survivors often reveal unbelievable quirks of fate in their experiences, before they came out of the Nazi hell. I shall adduce only a few examples. The destiny of people transported to the extermination camps was usually sealed as soon as they walked in, by a German officer's pointing left or right . Those directed to the left went to the gas chambers, to be killed and then burnt in the crematoria; those directed to the right went to a labour camp. In general able-bodied men and women were sent to camps, but if an able-bodied person happened, say, to hold a child in order to help a fellow victim walk with several other children, that person would also be directed to the left. There is also the well-known testimony of Anna Duklauer, one of three hundred women routed by mistake on a train to the Auschwitz extermination camp, who were taken to the gas chamber but saved by Oscar Schindler, the Righteous Gentile, just before their gassing: he claimed them as workers allotted to his factory. The women released on that occasion constituted the only shipment to come out of Auschwitz during the Second World War (Duklauer, 2007).

Prisoners in the labour camps had some chance of survival. Let me bring here the scene described by Zalmen Gradowsky, a

Jewish member of a Sonderkommando, who was killed in Auschwitz (the Sonderkommando units were made up of inmates who had to dispose of the corpses of the murdered, and after a period they were killed too). In his notes/diary, which were found under the ashes of a crematorium, he describes the morning roll call: 'At almost each block, beside the men standing in line, bodies of persons were lying. These are the victims of the night that have not lived to see the day. Even yesterday they were standing numbers at the roll call and today they lie, lifeless and motionless.' This kind of scene was depicted in the drawings of Jan Komski, a Polish Catholic who had himself been an inmate at Auschwitz, but survived (Gradowsky, 1973; Komski, 2006; Patterson, 1999). Two of the rebellions which took place in labour camps – those in Sobibor and Treblinka – were successful and some of the prisoners either ran away or survived, but the rebellion in Auschwitz-Birkenau was not successful: the insurgents managed to blow up one crematorium, but all of them were killed. The 'final solution' was the Nazi project of systematic annihilation of all Jews, and it was executed to the point where only a minority of Europe's Jews survived; their largest part was exterminated. Whatever the ultimate destiny of Jews, the Jewish people became, as Annette Wieviorka put it, 'a community defined by fate' (Wieviorka 1994).

Of course, this notion of 'fate' can be applied to natural disasters and everyday occurrences as much as to humanly created destruction. It makes no difference to the victims how their fate came about: was it through the dreaded Spanish influenza or through the bubonic plague, through the horrible consequences of wars, through a devastating tsunami or through some horrendous terrorist attack? In situations of this sort, anyone can be swept off the face of the earth in an instant: the victim's 'fate' can befall any individual. This brings it home that our world is full of the unanticipated, the precarious, the unknown. Does it also make us see our existence as being based on, and surrounded by, disorder, chance and chaos? But surely, there is also a great deal of evidence concerning the basically orderly nature of our existence. Constant rules, set conditions and uniformity do exist. It is to these evidently contradictory facts and aspects, and to the very notions which lie behind such contradictions, that we shall turn below. But first we shall consider Nicholas Rescher's brilliant treatise on 'luck', which also throws light on the notions of 'fate' and 'uncertainty'.

Rescher's World of 'Luck'

Nicholas Rescher examines most creatively, and from every possible angle, the phenomenon of luck, providing us with a thorough analysis and a comprehensive account of this phenomenon, its sources, and what it means for the human condition. Just to exemplify briefly and vividly what luck is, here are two quotations from Rescher: 'a small gust of wind can make a difference between a bullet's hitting or missing'; and, 'a small twinge, can cause a person to misstep and slip into the way of an onrushing car'. In his introduction to the subject, he makes the following basic statement:

> The human significance of luck stems from the fact that it is one of the characteristic factors that define our condition. For while we are intelligent agents who make our way by thought along the pathways of a difficult world, we are agents of *limited* knowledge who do and must make our decisions in the light of incomplete information. And for this reason we are inevitably at the mercy of luck. Our choices and decisions propose, but the ultimate disposition is at the mercy of a force beyond the limits of our cognitive and practical control. (Rescher, 1995)

It is somewhat surprising that Rescher refers to luck as 'a force', 'a rogue force', and that '[l]uck is the determiner of much of what happens to us in this world', although he does say that its existence is due to the power of chance, chaos and choice. Yet the first impression one gains is that luck produces some outcome. But, surely, luck *is* the outcome; it is a product of, not a producer of, some of the problematic conditions and circumstances in which we live. Later in his treatise, Rescher appears to be correcting himself when he states: 'Luck is not a force, factor or agency of some sort. We would be ill advised to reify – let alone personify – it. It is simply a reflection of a fact of life.' We live in a world which chance, chaos, choice, and ignorance make 'unamenable to prediction and [which] . . . thereby creates room for luck' (Rescher, 1995). Therefore luck is something unpredictable, which just happens, and something which has an evaluative status – that is, its result can be good or bad.

What interests me in particular is the state of unpredictability analysed by Rescher. As he says, 'luck's presence is pervasive in human affairs because our predictive capabilities are decidedly limited', and the 'natural impediments to prediction obtain in so far as the future is *developmentally* open – causally undetermined or underdetermined by the existing realities of the present and open to the contingencies of chance and choice'. Rescher sees luck and chance as two sides of the same coin, and differentiates between luck and fate, regarding the latter as missing the element of surprise, of chanciness and unpredictability. As mentioned above, there are four factors that bring about unpredictability: chance, chaos, choice, and ignorance. Out of the four, chance and chaos 'are in substantial measure genuinely random in not conforming to any sort of definite, outcome-determinative rule'. 'They are actually orders of a sort, albeit orders of a certain highly irregular sort.' The processes where chaos is at work are to be found in systems 'so highly sensitive to conditions that very minute differences in an initial state can nevertheless engender very great differences in result'. The two factors just mentioned represent nature's inbuilt characteristics – hence nature's capacity to cause unpredictability – while the other two factors are the making of human agents. Both choice and ignorance are internal to human beings, linked to their wishes, calculations, knowledge, and so on. There is no way of judging with any accuracy the way these factors will influence actions. It is in this sense that they add to unpredictability (Rescher, 1995).

The very fine distinction that Rescher makes between fate and luck seems to me a view difficult to hold and defend. Let us take his example of the discovery of 'a large but heretofore undetected meteor [which] is on a collision course with the earth' and which, through the consequences of its enormous destructiveness, would render our planet unable to sustain life any longer, and thus would seal humanity's fate. In Rescher's view, 'our extinction would (strictly speaking) be unfortunate rather than unlucky' (Rescher, 1995). One could, however, argue that the generation to which this catastrophe might occur would be totally unlucky to meet such a fate. I have in fact shown above, in one of my examples, how fateful events related to collectivities contain cases of individuals with good or bad luck. I am referring to the Holocaust, which sealed the fate of the vast proportion of European Jews and, as we have seen, the few who survived usually did so through individual luck. Fate and luck are intricately bound up. Even the notion of unpredictability does not always hold, at least not entirely, where events appear to be dictated by fate or luck. Take the instance of those who become ill as a result of smoking or consuming unhealthy foods: are such people simply unlucky, since others with the same consumption patterns do not become ill? Or, if a driver under the influence of drugs and alcohol escapes injury in an accident whereas another meets his death in it, are they simply lucky or unlucky? Such questions show the complexity of those – numerous – situations in which a great deal of predictability does obtain so long as one considers a collectivity, but not its individual members: logically (or in principle), the same predictability should also apply to individuals within the collectivity in question, yet in practice individual outcomes are less predictable. True, a probabilistic prediction can be provided, giving the percentage of those who will contract a disease or die in a traffic accident out of the whole population considered. However, at the individual level the prediction can be no more specific than yes or no – or, in Rescher's terms, lucky or unlucky. To be among the ones included, or not included, in the predicted percentage is a matter of luck.

Yet another example will suffice to show that, even where events are supposed to be predictable, the course leading up to such events may change, and hence the ultimate outcome may be said, in Rescher's terms, to have been after all unpredictable. The events leading up to the Second World War make, unfortunately, a good illustration of this principle. With Adolf Hitler's ascent to power in Germany it became pretty clear, by the mid-1930s, that a war was inevitable. Yet Britain's leader, Neville Chamberlain, perhaps wishing to gain time, worked for a so-called peaceful solution. The closest ally, the United States, was also biding its time. It was not until Winston Churchill became prime minister in 1940, and the Japanese attack on Pearl Harbor in 1941, that action against Nazi Germany became the new reality. In other words, the war itself was predictable, but the events leading up to it, the line-up of the opposing sides, and its final outcome, remained under the cloud of unpredictability and uncertainty.

What do unpredictability and, more generally, uncertainty mean for the human condition? Although there are some patterns which are common across the physical, the organic and the human–social conditions – for instance both order and disorder are apparent in all three; all three have systems which are open to change; and in all three chance and chaos occur (Krausz, 1987) – I shall refer only to the way Rescher sees the specific conditions that determine human life:

> One of the most fundamental features of the human condition is exactly [the] circumstance of our finitude – the fact that we are limited creatures who have a decidedly limited cognitive and practical control over our fate and our future – creatures at the mercy of developments largely outside the sphere not only of control but also of knowledge and even reasonable expectation. (Rescher, 1995)

The limited cognitive and practical control that Rescher highlights reverberates in our lives, in particular in the form of maladjustments which human beings experience in changing conditions – for instance global warming, terrorist insurgence, pollution, and many other seemingly intractable problems. The latter continue to aid and multiply the chanciness and riskiness of life. Some of these problems may be solved and even eliminated, only to be replaced by other, equally chance-ridden, conditions. In Rescher's view, only prudence and a better understanding of the nature of things in our surroundings can accommodate us more effectively to our conditions. Rescher's conclusion is that predictability is a worthy aim, but it can never be fully attained – a state of affairs which leaves us in the suspense of uncertainty (Rescher, 1995).

Living with the Unknown

Given that much of human life is punctuated with uncertainty, humanity has been driven by a 'quest for certainty' and by the wish to 'escape from peril'. John Dewey points out that, in attempting to succeed in these endeavours, philosophers have clearly distinguished between practical doings, which were seen as being subjected to riskiness and uncertainty, and pure rational thought, through which it was envisaged that the certainty of science would triumph. The idea was somewhat like this:

> To these two realms belong two sorts of knowledge. One of them is alone knowledge in the full sense, *science*. This has a rational, necessary and unchanging form. It is *certain*. The other, dealing with change, is belief or opinion, empirical and particular; it is contingent, a matter of probability, not of certainty. (Dewey, 1930)

But this triumph of rationality and perfect knowledge was not to be a realistic aim. Science developed in the twentieth century in a different direction – that of indeterminism.

The quest for certainty by means of rationality was essentially the basic tenet of 'modern thought'. Stephen Toulmin shows in effect that this foundation of modernity was a misconception since, however much one may hope to attain 'certainty and clarity in theory', one must balance this aspiration with 'the impossibility of avoiding uncertainty and ambiguity in practice' (Toulmin, 1990). Furthermore, the very idea of modernity that there had to be a new 'scratch line', a 'new start' which would sweep away all that preceded was an unfulfillable dream; 'the belief that, by cutting ourselves off from the inherited ideas of our cultures, we can "clean the slate" and make a fresh start, is as illusory as the hope for a comprehensive system of theory that is capable of giving us timeless certainty and coherence'. All we can do, according to Toulmin, is to start from where we are, to use discriminatingly and critically the ideas available to us in our current and local situation, and to consider 'the evidence of our experience' (Toulmin, 1990).

There is ample evidence, as seen above, that our lives are full of uncertainties, which produce indeterminacy in science and a lack of control over events that affect our everyday lives. It is also obvious that there are limits to human knowledge, which creates a serious obstacle to predictability. Moreover, it is clear that the reality which surrounds us contains the elements of chance, riskiness and chaos, and hence our choices cannot be fully rational. Furthermore, the advances of science and technology that gave us nuclear power could, in a world of chance and of imperfect control, bring our civilization to an end – or to another primitive beginning. Is such a truth to plunge us into scepticism? And yet another question must be raised: how can we reconcile evidence of this nature with the equally widespread testimonials about the existing laws and rhythms of nature and about the fantastic genius that man evinces? Humanity has shown that it has been able to survive despite the risks and uncertainties in our world, by taking precautionary measures, by making great inventions, and by using its own astuteness. True, these abilities have not always been sufficient and catastrophes, wars and epidemics have taken their toll. Yet in our days the advances of science and human ingenuity are sustaining populations tenfold greater than in former generations and times, which leaves room for a positive outlook. But we still have to reconcile the apparent contradictions. The answer lies in nature itself. Chaotic systems intermesh with perfectly ordered systems – there being intermittency between systems (Peat, 2002). Again, chance is effectively at work and intricately combines with the rational element in the choices that human beings make concerning their actions day-in day-out (see Chapter 5). Since the conditions in which we live are thus reconcilable, our conclusion must be that humankind is not doomed, despite the all-pervasive uncertainty of our existence. Its destiny appears to be one of continuing to live with the unknown.

Bibliography

Barrett, A. M., 2005. 'Is It Alzheimer Disease or Something Else? 10 Disorders that May Feature Impaired Memory and Cognition', *Postgrad Med*, 117 (5) (May): 47–53, accessed 25.04.07 at www.pubmed.gov.

Blatt, Thomas (Toivi), 2004. *The Forgotten Revolt-A Survivor's Report*, Issaquah, WA: Thomas Blatt.

Bohr, Niels, 1958. *Atomic Physics and Human Knowledge*, New York: John Wiley.

Bowler, John V. et al., 1998. 'Fallacies in the Pathological Confirmation of Alzheimer's Disease', *Journal of Neurology, Neurosurgery and Psychiatry*, Vol. 64, No. 1 (January), pp. 18–24.

Cohen, John, 1964. *Behaviour in Uncertainty and Its Social Implications*, London: Allen and Unwin, pp. 11–13.

Davidson, Paul, 1994. *Post Keynesian Macroeconomic Theory*,

Aldershot, UK: Edward Elgar, p. 17 (quoted in Janeway 2006).

Dewey, John, 1930. *The Quest for Certainty – A Study of the Relation of Knowledge and Action*, London: George Allen and Unwin, p. 23.

Duklauer, Anna, 'Auschwitz', accessed 11.06.07 at http://www.auschwitz.dk/holoc/new_page_3.htm.

Gradowsky, Zalmen, 1973. 'Amidst a Nightmare of Crime' (original manuscript in Yiddish, translated into Polish, with an English abstract), Oswiecim: Auschwitz-Birkenau State Museum Publication.

Green, Marc, and John Senders, 2004. 'Human Error in Road Accidents', in *Visual Expert Human Factors*, accessed 22.04.07 at http://www.visualexpert.com/Resources/roadaccidents.html, p. 1.

Hacker, Jacob S., 2006. *The Great Risk Shift*, Oxford/New York: Oxford University Press.

Heisenberg, Werner, 1958. *The Physicist's Conception of Nature*, New York: Hutchinson.

Humphreys, Paul, 2000. 'Extending Ourselves', in M. Carrier, G. J. Massey and L.Ruetsche, eds, *Science at Century's End*, Pittsburgh/Konstanz: University of Pittsburgh Press/Universitätverlag, pp. 17, 18–21.

Janeway, William, 2006. 'Risk versus Uncertainty: Frank Knight's Brute Facts of Economic Life', accessed 14.05.07 at http://privatizationofrisk.ssrc.org/Janeway/printable.html.

Knight, Frank, H., 2002. *Risk, Uncertainty and Profit*, Washington, DC: Beard Books, pp. 224–5 (quoted in Janeway 2006).

Koliopoulos, Sophia, 2001. 'Parental Genetic Testing – Do You Really Want to Know Your Baby's Future?', in *The DNA Files – Parental Genetic Testing*, Soundvision Productions, http://www.dnafiles.org/about/pgm3/topic.html.

Komski, Jan, 2006. 'Brotherhood in Suffering', exhibit (drawings) held at the Florida Holocaust Museum, St Petersburg, Florida.

Krausz, Ernest, 1987. 'Physics of Disordered Systems', *Philosophical Magazine B*, Vol. 56, No. 6, p. 659.

Kruse, R., E. Schwecke and J. Heinsohn, 1991. *Uncertainty and Vagueness in Knowledge Based Systems*, Berlin: Springer-Verlag, p. 3.

Patterson, David, 1999. *Along the Edge of Annihilation*, Seattle: University of Washington Press, p. 159.

Peat, David F., 2002. *From Certainty to Uncertainty*, Washington, DC: Joseph Henry Press, pp. 4, 5, 9, 15, 135.

Pollack, Henry N., 2003. *Uncertain Science . . . Uncertain World*, Cambridge: Cambridge University Press, pp. 105, 107–21, 122–3.

Reitsma, Hans, 2007. 'The STARD Initiative – Towards Complete and Accurate Reporting of Studies on Diagnostic Accuracy', accessed

25.04.07 at http://www.consort-statement.org/stardstatement/htm.

Rescher, Nicholas, 1995. *Luck: The Brilliant Randomness of Everyday Life*, Pittsburgh, PA: University of Pittsburgh Press, pp. 4, 46; 12, 17, 53–4; 29–30, 42, 50–6; 67, 177, 192.

Searle, John, R., 1992. *The Rediscovery of the Mind*, Cambridge, MA: The MIT Press, Chapter 4.

Toulmin, Stephen, 1990. *Cosmopolis – The Hidden Agenda of Modernity*, New York: The Free Press, pp. 175, 178–9.

Treat, J. R. et al., 1977. 'Tri-Level Study of the Causes of Traffic Accidents', Report No. DOT-HS-034—3-535 (TAC) (quoted in Marc Green and John Senders, 2004).

Wieviorka, Annette, 1994. 'Jewish Identity in the First Accounts by Extermination Camp Survivors from France', translated by Françoise Rosset, in A. Astro, ed., *Discourses of Jewish Identity in the Twentieth-Century France*, New Haven: Yale University Press (Yale French Studies 85), pp. 135–51 (accessed through JSTOR, 12.06.2007).

3 The Cosmos: From Evidence to Speculation

Science is an enterprise which is never satisfied with existing knowledge. It is, therefore, continually producing new hypotheses, the testing of which requires hard evidence. This task is particularly daunting in a science such as cosmology, which studies the nature, origin and history of the universe. Here evidence must be sought from the very distant reaches of the universe, both spatially and temporally.

Milton Munitz describes the universe as: 'the comprehensive whole that includes all physical objects, processes and events within its scope'. Following Paul Davies, we can refer to the universe as an all-embracing concept of a complex, coherent, unified and consistent whole, displaying certain regularities and forces which follow definite physical laws. Munitz also enumerates a number of basic questions asked in recent scientific cosmology. One such question is whether the universe had a beginning, or whether it was always in existence; another question is whether the universe has a spatial limit – or is it of infinite spatial extent? (Munitz, 1986; Davies, 1992). In this chapter I shall discuss the answers cosmology has provided to these questions and the extent to which these answers are complete and buttressed by scientific evidence.

In order to answer such questions, the cosmologist propounds a theoretical model which has to confront the observational data, leading to an interplay between the empirical and its conceptual interpretation. This is the generally accepted scientific process. The data available to the cosmologist represent information he receives 'in the form of signals conveyed by electromagnetic radiation of various frequencies . . . obtained by means of optical and radio telescopes, infrared detectors, radio antennas, and satellite mounted telescopes'. To give some idea of how far into space and time advanced data-gathering instruments reach, let us note that radio-telescopes have probed enormous distances of billions of light-years, a light-year being 'the distance the light, travelling at the rate of 300,000 kilometers per second, traverses in a year'. The vastness of the universe may be comprehended if we think of the giant clusters of galaxies of which it is composed, a galaxy being a collection of stars held together, so to speak, by gravitational attraction. The sun is such a star, one among many stars in our own galaxy, the Milky Way, which is itself a member of a local group of galaxies. Our own group of galaxies is a relatively small cluster, numbering approximately twenty galaxies – some clusters have thousands of member galaxies. Such clusters or super-systems have also been described as gigantic 'cloud complexes' of galaxies (de Vaucouleurs, 1957). The galaxies within these vast clusters or super-systems are separated from each other by enormous distances, and the size of a cluster may be expressed in terms of millions of years that light takes to traverse it, the largest clusters being estimated at 100 million light-years across.

In what ways is this vastness, which we call the *universe*, observable and intelligible from a scientific point of view? Galactic astronomy and astro-physics have certainly widened our horizons and deepened our scientific understanding of certain aspects of the universe. Yet, as we shall see, the basic questions of its origin, its extent and its future remain unanswered by science. Since evidence is our main scientific criterion, let me enumerate, in the simplest terms, some of the major steps along the road to a better understanding of the cosmos. The idea that the universe is composed of many galaxies like our own was already proposed by Immanuel Kant in 1755. The idea became generally accepted by the beginning of the nineteenth century. But it was not until 1923 that it became scientifically established, through evidence which the US astronomer Edwin Hubble was able to provide with his new powerful telescope. By 1929 the Newtonian view of the universe as infinite and unchanging had to be abandoned once Edwin Hubble discovered that, in the spectrum of light emitted by the stars of the galaxies, there was a shift towards red. To understand the meaning of this discovery we must refer to the Doppler effect. In

1842 J. C. Doppler, a Prague mathematician, described a phenomenon that is true of all transmissions of energy. The wavelike characteristics of sound and light which are emitted from a moving source undergo a shift in their frequency relative to a stationary receiver. Thus an ambulance siren becomes shriller as it comes nearer to where we stand and it becomes less high-pitched as it moves away, since in our vicinity the sound wave travels a shorter distance to us and thus at an increased frequency, which results in the high-pitched sound of the ambulance siren. The same is true of light waves. A spectrometer picking up the different colour bands of the light waves will show that the light emitted by an object moving towards us shifts to the blue end of the color spectrum, the blue representing a shorter wavelength, hence a higher frequency. On the other hand, light emitted from an object moving away from us shifts to the red, that is, to a longer wavelength representing a lower frequency. The shift towards red could thus mean only one thing: the galaxies were moving away at velocities proportional to their present distance from us. That is, the farther away the galaxy is, the faster it is moving away (Munitz, 1986; Munitz, 1981; Schroeder, 1990). The conclusion, based on this evidence, was clear: our universe is expanding.

Now, with this changing and expanding universe astro-physicists had to propose a theory, to be upheld by evidence, as to how it all happened. Two contending ideas were put forward: one was the steady-state theory proposed by H. Bondi and T. Gold in the 1940s. They maintained that the universe always existed and will always exist, since new galaxies were continually forming, replacing the stars that consume their fuel, cool and die; there was new matter continually created which enabled the universe to renew itself. The theory does not state out of where came the new matter. There is no such evidence available (Hawking, 1989; Schroeder, 1990).

The competing theory was that of the Hot Big Bang, now called the Standard Cosmological Model. It was Georges Lemaître who first proposed, in 1927, that 'the universe began with an explosion of a highly condensed primeval "atom"'. His idea was developed and elaborated upon by George Gamow in the 1940s (Munitz, 1986; Domb, 1995). N. Aviezer provides the following brief description of Gamow's theory:

> According to the Gamow theory *the universe began 15 billion years ago*. At that time . . . there suddenly appeared *out of nothing* an enormous source of energy, called the primeval fireball, popularly known as the 'big bang' . . . The sudden appearance of the primeval fireball marked *the beginning of the universe* in the sense that before the big bang nothing at all existed. Thus, the big bang is the precise realization of creation *ex nihilo*. (Aviezer, 1990)

Despite the difficulty of explaining *scientifically* how creation out of nothing is possible, in the case of this theory there *is* evidence that such an event took place. The theory predicted that the event of the 'big bang' would still have traces today in the form of weak electromagnetic radiation coming from all over space. Confirmatory evidence of this prediction was hit upon in 1964 by two American scientists, Arno Penzias and Robert Wilson, at the Bell Telephone Research Laboratories. By the 1960s radiation detectors were tremendously improved; Penzias and Wilson

> were using an extremely sensitive antenna to measure galactic radio waves. While testing their antenna they observed some very weak unexpected electro-magnetic radiation that seemed to be coming simultaneously from all directions in outer space. It was soon realized that this radiation was precisely what was predicted by the big-bang theory. (Aviezer, 1990)

This cosmological discovery of the *isotropic microwave radiation background* – 'that is, a radiation background that is constant and equal in all directions of the universe' – was of great significance in upholding the theory concerning the origin of the universe (Schroeder, 1990). Not all scientists have accepted the Big Bang Theory. The Swedish scientist Hannes Alfven, winner of the 1970 Nobel Prize for physics, was best known for debunking the Big Bang Theory. Some scientists have pointed out how uncanny a resemblance there is between the Big Bang Theory and the Bible account of the creation, including the commentaries of some of the best known Jewish Talmudic scholars of a thousand years ago (Schroeder, 1990; Munitz, 1986).

The Universe Observed

Many facets of science have converged and aided the cosmological explanations of the structure and origin of the universe. Mathematical physics and observational astronomy and the application of the general relativity theory have given us a description of the large-scale space–time structure of the universe. The Standard Cosmological Model itself, an evolutionary model which is accepted by most scientists today, has established the idea of an expanding universe: the galaxies are rushing away from each other due to the initial explosion, the distant ones at a faster pace than the nearer ones. On the assumption that all galaxies have the same intrinsic luminosity – the apparent brightness of a galaxy can be taken as a measure of its distance – it is possible to measure the velocity of heavenly bodies through their emission of electromagnetic waves. It is also possible to know the chemical composition of the universe: for instance, the relative amounts of hydrogen and helium that are observed to be present in the universe are in accord with the theory. Each chemical element emits its own unique and complex spectrum of light, which enables it to be identified; as a result, it is possible to assess accurately the abundance of the various chemical elements in the universe. There have also been fresh insights into the way in which various temperatures and energy conditions played their role in the early stages of the development of the universe (Munitz, 1986; Hey and Walters, 1987; Domb, 1995; Aviezer, 1990).

Cosmological scientific knowledge of the kind just mentioned is based on observed evidence, although we must remember that 'the observed universe is only a very small fraction of the universe as a whole' (Munitz, 1986). Let us take an important scientific probe like the unmanned US spacecraft Galileo launched in 1989, which entered the planet Jupiter's atmosphere in 1995, sending back, for less than an hour, data on temperature, winds, ammonia clouds and other conditions. The data could prove of great significance for our knowledge of how the solar system was formed, and even provide insights into the original starting period of the universe. Yet in observational terms the tiniest imaginable area, over a minuscule time period, was explored in the context of the universe as a whole.

Milton Munitz raises the question as to what are the horizons or limits of the 'observable universe'. He points out that this depends on such major factors as: 'the situation of the observer; available instrumentation; and types of physical theory used in interpreting the observable data, including the conception of the universe as a maximally intelligible system to which the observable universe is related'. This means that the horizons of the observable universe are movable, flexible. However, there are certain unverifiable assumptions and certain facts which put limits on observability. First, the cosmologist is earthbound and confined to our galaxy, although information does reach him from beyond his galaxy. For instance, he learns of the isotropic distribution of galaxies (the uniformity observed in the *galactic* structure of the universe) and of the microwave background radiation through direct observational data. Yet the assumption that similar isotropic observations would be made by all other observers at different spatial locations is a hypothesis that is not directly verifiable. To continue with Munitz:

> Not only . . . is the cosmologist rooted in space, he is also rooted in time. Opportunities for observation are severely limited by the duration of an individual person's lifetime, as well as that of the whole scientific community. The time during which such information gathering has taken place or is likely to take place is minuscule compared to the large-scale space–time extent of the universe insofar as this is currently understood. In short, whether taken individually or collectively, the human observer occupies what is for all practical purposes a fixed point in space–time. There are no realistic opportunities for stretching the observer's opportunity for direct observation to match the entire space–time extent of the universe. (Munitz, 1986)

Another aspect of the severe limitations imposed on the cosmologist is that he can have very little knowledge as to where he is located in the universe. He knows that he is in the solar system and in the Milky Way galaxy, but as to where exactly *this* is located *vis-à-vis* other galaxies in the universe as a whole he has very little knowledge – and in certain ways no knowledge at all. Since he can observe the isotropy of the universe he may, at first instance, have some kind of notion or illusion that he is at the centre of the universe. But he will quickly give up the idea that our galaxy has any special central position, since by simple logic, 'if the universe is isotropic around us it ought to be isotropic about every typical galaxy', which also leads to the idea that 'if the universe is isotropic around every part, it is necessarily also homogeneous'. These logical arguments mean that the universe seems the same to all typical observers wherever they might be located, and that it looks the same in whichever direction the observer looks. So any observer who sees himself as if at the centre of the world-picture must realize that any other observer anywhere in the universe could see himself precisely the same way, at the centre of the world-picture (Trivers, 1985). These hypothetical propositions seem so natural that they have been called 'the Cosmological Principle' by the English astrophysicist E. A. Milne (Weinberg, 1983). In view of this principle it would make little sense to ask whether, compared to other galaxies, we are nearer to or further from the centre or the edge of the universe.

Despite the serious limitations mentioned, scientists using existing cosmic data – observations made in particle physics laboratory experiments – and applying current theories have tried to tackle questions concerning the origin and pre-origin of the universe and its eventual fate. The ideas and possibilities are fascinating. The expansion of the universe goes on, apparently for ever. Since there are gravitational forces inherent in the mass and energy of the matter of which the universe is made up, these forces prevent it from eventually disappearing into nothing, fizzling out, so to speak. Yet the gravitational forces are not so strong as to make it collapse onto itself, which would result in a 'big crunch', signalling its end, to match the 'big bang' that produced it initially. This suggests that there is a fine gravitational balance at a 'crucial density' of matter and energy. In order to test this last possibility, scientists have had to take into account the mysterious dark matter which cannot be seen, but nevertheless seems to exert a gravitational influence on stars and galaxies (Gribbin, 1995; see also Schroeder, 1990). Although taking into account things not actually observed has been justified by most scientists, including the famous physicist Max Planck, who 'vigorously asserted the reality of the unseen world' (Perutz, 1989), it is still true that, to give satisfactory scientific answers, there is need for more accurate and solid evidence. So what we have at the moment concerning basic cosmic questions are scientific ideas which are partly speculative.

Once we pose even more difficult questions we truly enter the realm of speculation. Thus, is the universe spatially and temporally open or closed? Is there something beyond the universe? Has the universe any boundaries and, in that sense, can things disappear off the edge of the universe? Or can the matter of which the universe is composed disappear due to a space–time 'edge'? Is there something beyond the universe, or did anything exist before the universe appeared? Were there any previous events of a universal sort? Was there a vacuum, a pre-existing void, before the beginning of the universe? The questions are overlapping, annoying in a sense, and certainly troublesome for the scientist, whose answers are conjectural at best (Davies, 1977).

Scientists have two options when faced with such questions. They can follow Alan Guth, who wishes to escape the bounds of existing theories and provide 'speculations on the origin of matter, energy and entropy in the Universe' (Guth et al., 1983). Of course such scientists will be confronted with certain basic questions, which currently appear intractable, such as the *origin* of mass and gravity. Alternatively they can follow Stephen Hawking, starting scientific explanations with the *singularity* of the big bang, 'that crucial interface between total nothingness and the start of our universe' (Schroeder, 1990). Since the existing theory cannot handle questions that go beyond our present understanding of the history of the universe, scientists taking Hawking's option will have to agree with him in saying that,

> as is the case, we know only what has happened since the big bang, we could not determine what happened beforehand. As far as we are concerned events before the big bang can have no consequences, so they should not form part of a scientific model of the universe. We should therefore cut them out of the model and say that time had a beginning at the big bang. (Hawking, 1989)

Stephen Hawking thus sets theoretical limits to the possibilities of a scientific understanding of the origin of the universe. As we shall see below, Steven Weinberg, a prominent exponent of the big bang theory, points to some limitations which are due to lack of evidence.

Inventing Models

But, despite the limited evidence at hand, scientists have proposed explanations of the early development of the universe, its current state, and the prospect ahead. It is interesting to note that, before Hubble's evidence of the receding galaxies, and well before the Penzias–Wilson discovery of the background radiation of the universe, the Russian mathematician Alexandre Friedmann proposed in 1922 two types of model concerning the possible state and fate of the universe. One envisaged the universe as expanding to infinity, should the density of its matter be *below* a certain critical value; the other saw the universe as finite, should its matter have a density *greater than* the critical value, thus producing a gravitational field strong enough to make it implode unto itself (Weinberg, 1983). The first model suggests a universe open and infinite, with galaxies escaping to infinity; while the second model would be a closed and finite universe, eventually falling back into a supremely dense form.

Today cosmologists seem to favour an equilibrium model, such as the Standard Cosmological Model, which we already described above. On the basis of current knowledge, the following brief history of the universe may be given, so as to arrive at an equilibrium model. Initially the universe was in a state of extraordinarily high temperature and infinite density. Then, at the end of the first three minutes, after the big primeval explosion, nucleosynthesis set in. This is the production, on a cosmic scale, of a mixture of all types of chemical elements from one or two simple types of atomic nuclei, like helium and hydrogen, through a process of large-scale nuclear reactions, of the kind still going on in the sun and other stars. After a relatively long time, once cooling set in, with continuing expansion, the elementary particles were able to clump together into stars and galaxies. The original explosion filled the expanding universe with energy and radiation, the size of the universe continually changing (a fact which is expressed in its continually expanding horizon): an unbounded sphere. However, the cooling process seems to have pushed the universe towards a thermal equilibrium; and the gravitational force, emanating from the matter of which the universe is composed, seems gradually to slow down the velocities involved in the expansion, and thus to hold back its expansion. But the gravitational force is too weak to make the universe implode.

Is such a model representing a universe which is truly in equilibrium or just in near equilibrium? Is it possible that it is representing a universe which eventually will come to some kind of halt? Nobody really knows the answers to such questions. No definite predictions can be made, since current evidence is insufficient. Steven Weinberg, for instance, admits that a great deal of uncertainty hangs over the Standard Cosmological Model and that, when it comes to specifics, our knowledge is limited. Thus he says: 'No one knows precisely what velocity our galaxy has relative to the cosmic distribution of typical galaxies, but *presumably* it moves at a few kilometres per second in some direction' (emphasis added). The understanding of the first few minutes of the development of the universe is essential for the Standard Cosmological Model; yet, as he states, although all the calculations are based on the Cosmological Principle – the assumption that the universe is homogeneous and isotropic – there is no evidence that this principle was valid at the earlier times of the universe's existence (Weinberg, 1983). Weinberg admits that the theories concerning the early universe are highly speculative; that the Standard Cosmological Model cannot be claimed to be true; and that it certainly cannot provide an unequivocal answer regarding the future of the universe (ibid.).

Given the uncertainties surrounding the Standard Cosmological Model, we may as well indulge in some more fanciful speculation. Howard Trivers refers to one possible cosmic scenario, namely one in which the universe would approach a state of accelerated contraction, ending up in infinite density. In such a case, he says,

> Some cosmologists believe that a kind of cosmic 'bounce' will take place and the universe will begin again to expand . . . If the universe does re-expand, its expansion will again be halted and followed by another contraction, followed by another bounce, and so on for ever

> . . . Were this correct, our universe is an oscillating universe, with an endless cycle of expansion and contraction stretching into the infinite past and toward the infinite future. Hence there would have been no beginning and there will not be an end. Thus expansion and contraction constitute the macroscopic rhythm of the universe. (Trivers, 1985)

As Steven Weinberg says: 'Some cosmologists are philosophically attracted to the oscillating model, especially because, like the steady-state model, it nicely avoids the problem of Genesis' (Weinberg, 1983), and I would add: it also avoids the question of the 'end' of the universe. But, as Paul Davies says, the idea of a 'cosmic bounce' – together with the idea that extinction is avoided by the 'bounce', which occurs because of quantum effects – are purely a conjecture. And 'the question naturally arises as to what condition the universe was in *before* the bounce occurred' (Davies, 1977).

It is quite remarkable to compare these 'modern' ideas, emanating from the scientific field, with beliefs clearly stated in the classical texts of the Midrash, which contains Talmudic and mystical commentaries on Judaism's oral law – commentaries which have their roots in antiquity. According to these texts, God created numerous worlds, and 'went on creating worlds and destroying them . . . until he created this one' (*Midrash Rabba*, 1939). Of course, some of the basic questions asked in recent scientific cosmology were debated by famous philosophers and astronomers in ancient civilisations and right through the Middle Ages.

The Cosmos and Human Existence

Some cosmologists wish to go beyond the kind of scientific models we presented and propose that we introduce *the anthropic principle* concerning the universe and the human existence it contains. This principle states that the very existence of human life in our universe depends on certain cosmological circumstances; but that, equally, the very knowledge of the universe depends on the existence of the conscious human observer. To put it another way, although one can assume that there exists a universe which has a set of determinate properties independently of human observation, and that such a universe can exist without human beings, the universe must contain life if it is to be observed and known. As Milton Munitz puts it, 'The universe and human life are coupled', and as a result we must move in both directions to understand the relationship – from the universe to man and from man to the universe, 'since they are mutually involved in a very special way'. This mutuality is stressed by Brandon Carter, who introduced the expression 'anthropic principle', meaning that 'what we can expect to observe must be restricted to the conditions necessary for our presence as observers'. Our human existence 'imposes a *selection effect* on what we could expect to observe with respect to the universe' (Munitz, 1986). To to put it another way, 'what we see is biased in favour of those parts of the universe where we can exist' (Drees, 1994). This particular relationship between the universe and human life means that the universe, if it is to be cognizable, must have human life; that the presence of human beings 'explains' certain physical constants in the universe, for without those constants life would not exist; and that the constants uncovered by science depend on the selective observation and understanding of human beings. Here it is interesting to refer to certain difficulties encountered in quantum mechanics that were pointed to by A. Einstein, B. Podolsky and N. Rosen in a paper in 1935, which came to be known as the EPR paradox. Briefly stated, they demonstrated that in quantum mechanics the description of reality cannot be complete, so that a probabilistic interpretation is necessary because of unknown hidden variables (van Fraassen, 1991; Sakurai, 1985). We must distinguish between the two theories: succinctly put, the EPR paradox points to the limitations imposed by empirical reality on observability, whilst the anthropic principle highlights the dependence of observable reality on human beings.

But the most interesting upshot of the anthropic principle is that the physical constants of the universe are so finely tuned that any change in one constant would affect the entire structure and ongoing processes, which as a result would mean that life as we know it would not exist. And, as Paul Davies points out, 'the fact that even slight changes to the way things are might render the universe unobservable is surely a fact of deep significance' (Davies,

1992). To elaborate on this a little, let me quote a prominent physicist, Moshe Kaveh, who in a newspaper article for the general readership puts it as follows:

> Human life on earth depends on the sun; the energy that life on earth receives from the sun is the result of a process of atomic reactions taking place within the sun; these reactions work according to certain physical laws which provide the sun's stability; should these change even just by the tiniest measure, the sun would explode. Again, if the orbit of our planet was, just by a few percent, nearer the sun, or further away from it, human life would not be possible since water, which is vital for life, would either completely evaporate or become deeply frozen; our orbit, one of many possible orbits, is therefore one that is particularly fortuitous and suited for life on earth. (Kaveh, 1994; see also 't Hooft, 1994)

To give another example, in the 1960s Fred Hoyle noted that carbon, which is crucial for terrestrial life, is received from huge stars, which by unique processes produce carbon and release it throughout the universe; yet, as F. J. Dyson of the Institute for Advanced Study in Princeton emphasizes, the vastness of the interstellar spaces and consequently our enormous distance from stellar explosions saves us from this deadly cosmic radiation (Aviezer, 1990; Davies, 1992).

So where does the anthropic principle lead us to? It certainly makes us ask a number of questions. Thus: considering that the scientific knowledge we produce concerning the universe depends on existing observations and observability, and on existing intuitions and interpretations made by certain observers at some given time or times, could there be other observers at some points of time, or the same observers at other points of time, who could come up with different observations, visions, intuitions, interpretations, and so on? Does it all depend, therefore, on the conscious state of the observer, and does this mean that there may be a number of coexisting realities, each 'seen' or 'imaged' by a particular observer in a unique situation? Again, considering that the universe may not give up all its secrets, so to speak, and that, therefore, what *is* knowable may not all come forth in one go, is it possible that the universe we comprehend at any one time is only part of some larger super-universe or a multiplex universe transmitting several messages/signals simultaneously, which are picked up differently by various human observers? Finally, could there be observers in the universe, other than human beings on our planet, who 'see' or 'have knowledge' of a universe quite different from that which our scientists comprehend?

In sum, is the universe we think we know just one of a number of different universes possible to comprehend? As Paul Davies tells us, such questions are taken seriously by many physicists and philosophers (Davies, 1992). Thus, according to H. Everett and J. A. Wheeler, the 'relative state' interpretation of quantum theory, which deals with physical phenomena on the micro-scale of electrons and nuclei within atoms, leads to the idea that the possible alternative outcomes of measurement represent alternative, parallel or successive universes (Shapere, 1984). The same idea was expressed more recently by David Deutsch, a physicist at Oxford University: since in quantum mechanics 'objects do not exist in a single state but possess a range of possible states with particular probabilities', this means that 'there exist different universes corresponding to each possible state of affairs' (*Economist*, 1996). Of course, the 'many worlds' interpretations of quantum theory and mechanics, and the very questions posed above, effectively take us out of realistic science and place us unquestionably into the realm of speculation bordering on science fiction.

Testing the Limits of Science

There are many areas studied by science which show the limits reached due to lack of clear evidence or to the incompleteness of existing theories. Cosmology is one such area; it has the distinction of showing these limits on a spectacularly grand scale. To exemplify, if an explanation is to be given for the formation of galaxies, inhomogeneities and anisotropy (that is, a lack of uniformity and smoothness) must have developed after the actual creation of the universe – a universe which afterwards underwent homogenization and isotropization. That is to say, without inhomogeneity no gravitational collapse to form galaxies and stars could have occurred. But, even if we adduce the quantum theory, according to which fluctuations arise spontaneously, which in the early universe could have given rise to inhomogeneities or clumpiness in the sizes needed to form galaxies, the evidence for this phenomenon is unobservable, and there is no theory that can explain 'why the universe should have any special initial degree of inhomogeneity' (Weinberg, 1983). The problems of the origin and nature of galaxies and of the incomprehensibility of creation *ex nihilo* loom large in the uncertainties surrounding the science of cosmology (Shapere, 1984; Trivers, 1985; *McGraw-Hill 1995 Yearbook of Science and Technology*, 1994).

No doubt, creative imagination and scientific advances, which are unforeseeable, will add to our knowledge in the future (Shapere, 1984). Yet, as Steven Weinberg says, ignorance in the area of the early universe is still great, and progress is basically of a conceptual sort (Weinberg, 1983). It seems that the mix of some evidence and a good dose of speculation gives rise to interesting scientific concepts, and may eventually lead to even more fascinating theories. But the theories about the early universe will remain mere conjectures, since no scientist was there to see actually what happened! (Weinberg, 1983).

Bibliography

Aviezer, Nathan, 1990. *In the Beginning*, Hoboken, NJ: Ktav Publishing House, pp. 9, 13.

Curran, Noel, 1994. *The Logical Universe – The Real Universe*, Aldershot: Avebury.

Davies, Paul C. W., 1992. *The Mind of God*, New York: Simon and Schuster, pp. 196ff., 199, 200, 217.

Davies, Paul C. W., 1977. *Space and Time in the Modern Universe*, Cambridge: Cambridge University Press, pp. 164–6.

Davies, Paul C. W. and John Gribbin, 1992. *The Matter Myth*, Harmondsworth: Penguin.

de Vaucouleurs, Gerard, 1957. *Discovery of the Universe*, New York: Macmillan, p. 289.

Domb, Cyril, 1995. 'Torah and Science – Are They Converging?', *BDD Journal of Torah and Scholarship* (Bar-Ilan University Press), No. 1 (Summer): p. 36.

Drees, Willem B., 1994. 'Problems in debates about physics and religion', in Jan Hilgevoord (ed.), *Physics and Our View of the World*,

Cambridge: Cambridge University Press, p. 204.
Economist, 1996. Vol. 338, No. 7949, 20–26 January, p. 90.
Gribbin, John, 1995. 'Hot Dark Matter Loses Its Mystery', *New Scientist*, Vol. 145, No. 1964 (11 Febr.).
Guth, Allan H., 1983. 'Speculations on the Origin of Matter, Energy and Entropy in the Universe, in A. H. Guth, K. Huang and R. L. Jaffe, eds, *Asymptotic Realms of Physics*, Cambridge, MA: MIT Press.
Hawking, Stephen W., 1989. *A Brief History of Time*, New York: Bantam Books, pp. 50, 51.
Hey, Tony and Patrick Walters, 1987. *The Quantum Universe*, Cambridge: Cambridge University Press.
Kaveh, Moshe, 1994 (8 July). Article in Hebrew in Tel-Aviv newspaper *Hatzofe* .
Midrash Rabba, 1939. Translated by H. Freedman, Vol. I (Genesis), London: Soncino Press, pp. 23–4 (Chapter 3), and 64 (Chapter 9).
Munitz, Milton K., 1981. *Space, Time and Creation – Philosophical Aspects of Scientific Cosmology*. 2nd edn, New York: Dover Pulblications, Inc.
Munitz, Milton K., 1986. *Cosmic Understanding*, Princeton, NJ: Princeton University Press, pp. 6, 50, 116–17, 125, 150, 152–4.
Olby, R. C., G. N. Cantor, J. R. R. Christie, and M. J. S. Hodge , eds, 1990. *A Companion to the History of Modern Science*, London/New York: Routledge.
McGraw-Hill 1995 Yearbook of Science and Technology,1994. New York: McGraw-Hill Book Company, pp. 186–7.
Perutz, Max F., 1989. *Is Science Necessary?* London: Barry and Jenkins, p. 17.
Sakurai, J. J., 1985. *Modern Quantum Mechanics*, Menlo Park, CA: The Benjamin/Cummings Publishing Co., Inc., pp. 226–7.
Schroeder, Gerald L., 1990. *Genesis and the Big Bang*, New York: Bantam Books, pp. 24, 42, 71–3, 78–9, 83.
Shapere, Dudley, 1984. *Reason and the Search for Knowledge*,Dordrecht: D. Reidel, p. 176 and Chapters 9, 11.
't Hooft, Gerard, 1994. 'Questioning the Answers or Stumbling upon Good and Bad Theories of Everything', in Jan Hilgevoord, ed., *Physics and Our View of the World*, Cambridge: Cambridge University Press, p. 35.
Trivers, Howard, 1985. *The Rhythm of Being – A Study of Temporality*, New York: Philosophical Library, pp. 33, 39.
van Fraassen, Bas C., 1991. *Quantum Mechanics: An Empiricist View*, Oxford: Clarendon Press, pp. 338–9.
Weinberg, Steven, 1983. *The First Three Minutes*. Glasgow: London:

Fontana (Flamingo Series), pp. 30–3, 63, 76, 118–19, 143–5, 148, 160 (original publication London: André Deutsch 1977).
Weinberg, Steven, 1993. *Dreams of a Final Theory*, London/ New York: Hutchinson Radius/Random House Vintage Books.

4 The Individual Facing the World

There has been a great deal of philosophical debate as to how to conceptualize the 'individual'. Has every single human being a unique personality, which defines him, and only him, at the core? Does this personality change and develop throughout its existence at every age and stage? And, above all, how does such a personal identity relate to other human beings, all with their unique personalities, and more widely to society as a whole?

It is quite clear that, despite the basically common features of the biological and psychological make-up of all individual beings in the world, there are specificities that attach to each one of them which define their individualities. In the vast majority of cases no two people look alike, and for this reason we are always amazed at the phenomenon of identical twins. But even in that case closer examination will reveal that there are distinct and unique features, which distinguish between them.

Genetically, the clearest distinction is found in the DNA, the nucleic acid molecule, which contains the 'genetic instructions' for the development and functioning of all living organisms. At this level, each living creature, including the human being, is unique in that each one has a different DNA composition. And even in the case of identical twins, who do have the same DNA, the fingerprints are not identical, due to the developmental environment in the uterus before birth. Again, many identical twins share IQ and psychological profiles as well as EEG tracings (that is, the electrical activity in the brain), and much has been reported and written on what appears to be the uncanny connections and psychological bonding between identical twins. On the other hand, epigenetic modifications do take place during the lives of identical twins, which shows that some genetic differences can develop between them. Whilst there is no denying that genetics plays a most important role in the case of identical twins, this observation is also true, and perhaps to an equal extent, in the case of fraternal twins and ordinary siblings. By the same token, although to various degrees, the environmental influence of parenting is most significant for many aspects of the development of human characteristics and of behaviour in offspring. The importance of parenting has been specially highlighted by some researchers both for identical and for fraternal twins (Klein, 2003). It is certainly not my intention to enter the 'nature versus nurture' debate. My aim is simply to point to the special identity that each human being carries throughout his life, whatever the effects of heredity and environment might be. But, whilst each human individual does have a special identity, this is not to say that he is independent of other individuals. After all, he carries his genetic and cultural inheritance from past generations and, similarly, he leaves his own legacy to future generations.

The major question that arises is how the individual, with his specific personality, including his wishes, his will, his intelligence and his comprehension of events, fits in with the vast number of other individual personalities, who constitute the human world he lives in. In a preliminary fashion one may suggest that society and culture have leveling and unifying effects, by stressing the common aspects of life that apply to all individuals, and thus they enable every individual to fit into society as a social actor. I shall delve into this topic by reviewing the ideas of philosophers, sociologists and psychologists.

The Individual and his Self

The issue of how an individual relates to other fellow human beings and to the world around may be handled in the first place by establishing, with the preeminent philosopher Immanuel Kant, 'that even the simple fact of being self-conscious, being aware of oneself, assumes an objective world: "the mere consciousness of my own existence proves the existence of objects in space outside me"'

(Speake and Mitchell (eds), and Antony Flew, editorial consultant, 1979).

This awareness of one's self has led psychologists to study and theorize how awareness develops in early childhood and how it affects the individual's motivations and behaviour. It was Sigmund Freud, the father of psychoanalysis, who propounded the theory that the psyche – that is, the mental structure of a person – is divided into the unconscious, preconscious, and conscious; subsequently Freud put forward a

> structural categorization into id, ego, and superego. The id was defined in terms of the most primitive urges for gratification in the infant, urges dominated by the desire for pleasure through the relief of tension and cathexis of energy. Ruled by no laws of logic, indifferent to the demands of expediency, unconstrained by the resistance of external reality, the id is ruled by what Freud called the primary process directly expressing somatically generated instincts. Through the inevitable experience of frustration the infant learns to adapt itself to the exigencies of reality. The secondary process that results leads to the growth of the ego, which follows what Freud called the reality principle in contradistinction to the pleasure principle dominating the id. Here the need to delay gratification in the service of self-preservation is slowly learned in an effort to thwart the anxiety produced by unfulfilled desires. What Freud termed defense mechanisms are developed by the ego to deal with such conflicts. Repression is the most fundamental, but Freud also posited an entire repertoire of others, including reaction formation, isolation, undoing, denial, displacement and rationalization.
>
> The last component of Freud's trichotomy, the superego, develops from the internalization of society's moral commands through identification with parental dictates . . . Only partly conscious, the superego gains some of its punishing force by borrowing certain aggressive elements in the id, which are turned inward against the ego and produce feelings of guilt. But it is largely through the internalization of social norms that the superego is constituted, an acknowledgment that prevents psychoanalysis from conceptualizing the psyche in purely biologistic or individualistic terms. (*The New Encyclopaedia Britannica*, 1992, *s.v.* 'Freud')

The most important concept promoted by Freud, although others before him put forward similar concepts, was that of the unconscious part of the mind. That is to say, quite contrary to the common assumption that people have immediate and unproblematic access to their own minds – expressed for instance in René Descartes' dictum *Cogito ergo sum* ('I think, therefore I am') – Freud held that many central aspects of a person remain radically inaccessible to the conscious mind, and that this fact explains some of the behaviour of the individual which otherwise would be unexplainable. And this is where psychoanalysis enters the stage: psychoanalysis can drag out important information from the unconscious. Freud had a tremendous influence on western thought, but his theories about personality and the human mind remained controversial, especially as they did not stand up to the standards of positivist science.

A different psychological approach from that of Freud, but one equally concerned with the individual, was that of Alfred Adler, whose basic concept was that an individual's uniqueness resides in the totality of the individual in question, regarded as an undivided personality: we can only understand the various aspects of that personality through the total. The human being, Adler tells us, is constantly moving towards achieving his goal, which at the most basic level is expressed in the endeavour to survive. This basic goal has a starting point in early childhood, being the result of a combination of factors:

> With his inheritance and the thousandfold impressions given by his physique, his environment (people and surroundings), as well as the influences of climate, culture, and society, the child creates his own way of survival and development . . . It is with his creative ability that the child tries to find the way in an unknown world, in which he has to find his place and has to achieve significance. Out of what is innate and what is outside of him, the individual at an early age creates his personal goal, which from there on dictates his actions, thinking and feeling. Only if this personal goal is included in his concept of his significance can he become an integrated personality. (De Vries, 2006)

Adler emphasizes that the human individual is a social being; therefore too much individualization leads to isolation, and even neuroticism. An integrated personality, capable of reasoning and of using his creative ability, will see his possibilities of choice; this means that his life is not simply determined by outside factors. As a social being, the individual will cooperate with others and will adjust to conditions, including societal ones, in his striving to achieve his personal goal (de Vries, 2006; Adler, 1956).

Unlike Sigmund Freud, Alfred Adler and other psychologists following them, the behaviourist branch of psychology, whilst interested in the individual's observable acts, gave no importance to the inner personality. Behaviourists claim that the task of psychology is not to study the inner state, the 'consciousness' of the human being, since this is not observable. It is rather *the behaviour or activities of the human being* that must be studied. The category of behaviour includes all that human beings do and say, and the description of behaviour must be in terms of *stimulus and response* (Watson, 1925). The best known exponent of behaviourism is B. F. Skinner, who seeks to give a causal explanation of the behavioural patterns of human beings – an explanation which, he claims, is based on science, since it would not differ fundamentally from the explanations of behaviour in lower organisms (Skinner, 1953). The extreme behaviourist approach holds that psychology has no truck either with physiological processes in the human organism, for instance neural transmissions or chemical reactions in the body, or with mental processes such as invisible streams of an individual's conscious thoughts and memories. Only overt observable behaviour is to be taken into account (Krausz, 2000).

As behaviourist psychology lost ground, experimental psychology came into prominence in the second part of the twentieth century. Experimental psychology deals with the cognitive processes going on in the mind as well as with the behavioural aspects of personality: it studies perception, thinking, learning and memory, making experiments on individuals placed in controlled situations. For instance, experimental psychology has been researching individual differences in subjective experience such as the feeling of pain or of phantom limbs; individual differences in levels of awareness; or individual differences in subconsciousness and in self-consciousness

Developments in psychology which use the experimental

method and link the psychological with the physiological aspects of personality have tended to shift the discipline from the domain of social sciences to that of natural sciences. In this respect the major question that loomed, and still remains basic, is whether the mind–brain dualism is tenable, or whether the equivalence or synonymity between the two can be proved. In other words, does a separate 'system' exist by means of which we can think, know, remember, feel, hate, want, and so on, no matter how such aspects of what we call the mental component of our life may be facilitated by the brain, by the nervous system, and in fact by the whole human body? Alternatively, are these cognitive and volitional events (and their subjective nature) produced simply by the brain and by other material aspects of our biological make-up, as responses to the various processes in which our lives take part? There are some strong arguments for the unification of mind and brain: thus scanning techniques can observe 'brain activity during mental processes and infer the locality of certain functions'. The regions of the brain that enable us to see have been mapped in this way, and more recently researchers 'have identified a region responsible for an auditory process, [enabling us] to locate sounds even when the listener isn't concentrating on them' (Siegel-Itzkovich, 2007; Deouell et.al., 2007). Furthermore, case studies of brain-damaged patients have often isolated more specific areas of function. However, there are some counter-arguments. No specific locations have been identified where 'higher functions' are concerned – for instance solving complex problems or showing creativity and imagination. More pointedly, the question has been asked whether love can be defined physiologically, as a 'dynamic interplay of neural activity'. In more general terms, it could be argued that the brain activities observed and the inferences made concerning the brain's responsibility for certain functions could also be interpreted as showing 'the interplay between mind and brain rather than the sole work of the brain'. There seems to be no clear conclusion ('The Mind and Brain – Synonymous or Dichotomized', 2007).

Some of the best 'minds' of the twentieth century have struggled with 'the body–mind problem' by considering dualist and materialist stands. Some have pointed to a possible solution to be found in the relationship between brain and mind, assuming that their interaction is located in the brain (Popper and Eccles, 1977).

A central feature of the functioning of the mind–brain is consciousness. John Searle deals with the difficulty of giving a definition of 'consciousness'. He offers certain notions about it: for instance a person enters a state of consciousness when he is awake, or has a sense of awareness – although the latter 'is more closely connected to cognition, to knowledge, than is the general notion of consciousness'. The real difficulties arise when we consider Searle's view as to what consciousness is in a basic sense. On the one hand, he considers it to be a biological feature of the brain caused by neurobiological processes; on the other hand, he accepts that consciousness is an irreducible feature of physical reality. The most succinct and substantive statement of Searle is as follows:

> Conscious mental states and processes have a special feature not possessed by other natural phenomena, namely subjectivity. It is this feature of consciousness that makes its study so recalcitrant to the conventional methods of biological and psychological research and most puzzling to philosophical analysis. (Searle, 1992)

The subjectivity factor highlights, of course, the 'individuality' aspect of consciousness. The debate over what 'consciousness' is still goes on, and we do not as yet have a proper scientific understanding of its workings (Krausz, 2000). The mind–brain problem, and especially the feature of consciousness, are obviously significant when considering an individual's mental state, since the latter encapsulates self-awareness, attitudes and intentions concerning one's responses to the environment in which one lives.

The difficulties of examining and understanding scientifically the individual's total personality or identity do not prevent us from seeing the individual as a separate human being – not simply as a biological organism, but as a person with his own human identity. In order to understand what this human identity is, it is necessary for us to widen the horizon by considering the effects that life in a social context has on his personality, and to examine his relationship with other individuals – and in particular the way in which the individual and the social world react to each other and influence each other.

The Individual and Society

In considering the relationship between the individual and society, my starting point is 'the theory of social contract', propounded by seventeenth and eighteenth-century philosophers. The basic ideas of this theory emanate from Thomas Hobbes, John Locke and Jean-Jacques Rousseau. These philosophers held that, in the state of nature – that is, in the original, primitive conditions – people lived in anarchy, as Hobbes expressed it: 'a war of every man against every man'. There was no organized society. The disorder was intolerable, and therefore people realized that they had to surrender some of their individual natural rights and liberty, in return for an organized orderly state and society, under a monarch or a government. This is how the social contract was forged. Locke used this theory as a basis for advocating popular sovereignty, 'the idea that the monarch or government must reflect the will of the people'. In the same vein, Rousseau 'found the general will a means of establishing reciprocal rights and duties, privileges and responsibilities as a basis of a state'. The principles contained in this theory were similarly used to justify the American and French revolutions of the eighteenth century. Thomas Jefferson, for example – a political philosopher and the third president of the United States (1801–9) – 'held that the preservation of certain natural rights was an essential part of the social contract, and that "consent of the governed" was fundamental to any exercise of governmental power', *The Columbia Encyclopedia*, 2001–5, *s.v.* 'Jefferson, Thomas'; Barker, 1960).

Let us look more specifically at Jefferson's ideas concerning the rights of the individual. For Jefferson, it was a self-evident truth that man was born with certain 'natural rights', which were 'inherent and unalienable'. In this respect Jefferson followed John Locke's ideas of natural rights and emphasized the main point that 'subsumed under it [under natural rights] was freedom, with emphasis on individual freedom. This was the pivot around which everything else revolved' (Padover, 1965). Furthermore, he did not accept the totally negative Machiavellian or Hobbesian notions of the nature of human character. On the contrary, 'Jefferson constructed his whole political philosophy on trust in the character of man, both as an individual and as a member of society' (ibid.).

Jefferson's trust was based on the belief that the individual, 'in sum, was a moral and rational creature, able to differentiate between right and wrong, between justice and injustice. His moral instinct guided him in making decisions as to what was good or bad for him and for society' (Padover, 1965). This was the basis on which Jefferson became 'a champion of liberty'. This is clearly reflected in the United States Declaration of Independence of 4 July 1776, which he drafted, and which lays down every individual's natural right to life, to liberty and to the pursuit of happiness. In wider terms, Jefferson believed in, and highlighted, the principles of personal freedom and democratic rule, such as equality before the law, and he condemned government oppression. He championed the freedom of religion and called for education for all, and he expressed his animosity to slavery and racial discrimination. But what marred his libertarian record was the fact that he was a slave owner and he agreed, for political reasons, to delete the clause on the abolition of slavery from the declaration. Despite such lapses, he certainly contributed greatly to advancing the principle of individual freedom, both as a political leader and as a practical philosopher (McDonald, 1976; Koch, ed., 1971; Finkelman, 2001; 'The Declaration of Independence', 2007; Weymouth, ed., 1973).

Jefferson's emphasis on the rights of the individual can be contrasted with the Platonic ideas followed by the German philosopher G. W. F. Hegel (1770–1831), whose thought became popular in Europe. Hegel emphasized the supremacy of the collective, the nation, the state, as against the freedom of the individual. Especially in the twentieth century, such theories served dictators of various political strands to abrogate the rights of the individual totally, in the name of the higher values and interests of the state (Popper, 1966).

The idealistic notions emphasizing individualism lead us to the issue of 'free will' in human life. Can the individual, as a rational agent, exercise full control over his actions and decisions? Or, much to the contrary, are his actions and decisions determined by the circumstances and actions of others and by the restraint put on him by society or by divine disposition? I shall avoid this endless philosophical debate and return to what is a kind of mid-point – the compromise with which I started above –resuming in a critical way the discussion about 'the social contract'. The latter seems to provide a neat way out of the quandary, yet it is subject to serious criticism.

A basic criticism of the social contract is that of David Hume, the brilliant eighteenth-century Scottish philosopher and contemporary of Rousseau. Hume criticizes the social contract theory by pointing out that

> the benefits of society can be calculated only by those who already have *some experience* of it. Consequently it cannot have arisen as a result of such a calculation. For Hume, the idea that society could have arisen through a contract is simply a myth.

Such an idea lacks historical support and is incoherent; 'it attributes to individuals in a state of nature attitudes and procedures which presuppose the social life they are supposed to explain' (Mounce, 1999). In other words, if social life replaced the Hobbesian totally individualized and anarchic type of existence as a result of individuals calculating that social life was preferable, then such individuals must have had a notion of what social life was like.

The social contract theory has been challenged mainly at the level of its basic assumptions. Thus the first question to be faced is this: what historical and empirical evidence is there for its material presuppositions, such as that of the existence of an originally and naturally anarchic, and totally conflictual, world? Secondly, are the epistemological and methodological aspects of this theory well grounded? Thirdly, would a similar theory not be derivable from a different sociological premise, which avoids the 'original state of nature' idea? These are the critical points formulated by John Bewaji, who suggests, furthermore, that 'it is very presumptuous of the social contractarians . . . to suppose that they have captured **the** basis of social contract and, hence, provide *the only logically valid* account of "democratic practice" universally'. Such basic points of criticism have remained unanswered; yet, despite this fact, modern scholars have continued the tradition of contractarianism, albeit from a perspective adjusted to the contemporary world (Bewaji, 2007).

John Rawls, a particularly well-known scholar, shifts the social contract theory away from the 'original state of nature' premise by emphasizing the *legitimacy* of a collectively enforced social arrangement, which would arise out of the agreement by 'the people who are subject to it', so that all the parties to the agreement have 'reason to honor their responsibilities' – either because of the clear terms of the agreement or simply because it seems reasonable to do so. Michael Lessnoff calls such reasonings 'the *voluntaristic* and *rationalistic* readings of the contract'. To sum up: 'In its modern guises, contract approaches are not intended as accounts of the historical origins of current social arrangements, but instead as answers to, or frameworks for answering, questions about legitimacy and political obligation' (D'Agostino, 2003; see also Rawls, 1999 and Lessnoff, 1986).

I shall now leave the above – narrow – framework of the social contract idea, whatever its contribution might be to the understanding of the relationship between the individual and society, and I shall consider instead the far more balanced and wider view of this relationship, which was adopted by classic and modern sociologists and social psychologists. Such views will provide us with larger frameworks, which include many areas of sharing and cooperation, and which go beyond the merely basic idea of socialized life as it is exemplified by the notion of individuals being governed by a state – as social contract theorists emphasize.

Basic Sociological Views

There is a very rich sociological literature dealing with the topic of the relationship between the individual and society. It will be enlightening to take here just two of the best known classic sociologists on this issue and to contrast their ideas: these are Emile Durkheim and Max Weber.

For Durkheim, the key issue is to explain the underlying social order in terms of processes that promote the integration and solidarity of individuals. In all walks of life there are dominant institutional patterns – be it in marriage and the family, in the work environment, in the political arena, in cultural pursuits, in religious worship, and so on. Quite naturally, individuals are engaged in activities pertaining to all these areas, but the phenomena created are social and transcend the individual. The latter integrates into the social groups of which he is a part and expresses common beliefs and values, integration being based on primordial social bonds. The degrees of integration, however, vary. The best example Durkheim gives is that of suicide, an act which, on the face of it, is very much an individual affair. Not so, according to Durkheim. On the one hand, there is egoistic suicide, which is due to the individual's not having sufficiently strong ties with the social groups, as in the case of the not-married. The opposite is true of altruistic suicide, which is the result of an excessively high level of integration, usually culturally induced, where the individual becomes insignificant in his own eyes and is prepared to sacrifice himself for the public good. These individual acts, when aggregated, become social facts, expressed in statistical form as suicide rates. Similarly, there are social events such as religious ceremonies, in which many individuals participate in accordance with the rules of a given religious institution; but the collective event has its own social significance. Thus society both attracts the individual's sentiments and wish to be part of it, and regulates the individual's activities. In other words, Durkheim elevates society to a special expression of human life, although society is constituted by the acts and interrelationships of individuals. In essence, Durkheim regards social life as a *sine qua non* for the individual and for human life as such (Durkheim, 1952; Rex, 1961; Broom and Selznick, 1968; Coser, 1971; Ritzer, 1983).

Max Weber, too, examines social behaviour, which he defines as 'activities by the individual oriented towards others', and he stresses that it is the individual who has *subjectively intelligible orientations*. Weber does not ignore, however, the concept of collectivity – for instance in 'the legal' or 'the economic' spheres, which do influence the behaviour of the individual and provide alternative modes of social behaviour. However, Weber stresses the importance of the individual, showing that changes in society are brought about through individual actions and personal goals. In his well-known thesis on the rise of capitalism in western Europe, Weber pointed to a fact which seemed to him strange – namely that there was a correlation between the rise of Protestantism and the growth of a capitalist economy in certain parts of Europe. His explanation was that the Protestant religion and its new spirit, which spread throughout western Europe from the sixteenth century on, and especially its Calvinist doctrines, influenced greatly the behaviour of people. Protestantism encouraged the individual to regard work as a sacred calling which leads to salvation – the latter requiring toil, thrift and personal success. The actions of individuals, in their attempts to live up to such ideals, made in fact possible the saving of capital, which in turn enabled the advancement of economic development. In consequence, Weber saw religion as a major determining factor in social life. His approach was to stress the importance of individual actions and the meanings which individuals attach to their actions. Weber's emphasis on the importance of personal belief and action is clear. True, he admitted that Protestantism, and subsequently capitalism, were mass phenomena, widely binding all members of society; but he still stressed that these phenomena were largely the outcome of individual actions, which had to be seen in personal terms – whether as spiritual salvation in the earlier phase or as material interest in the later stage of capitalism (Weber, 1930; Rex, 1961; Bendix, 1966; Broom and Selznick, 1968; Coser, 1971; Ritzer, 1983).

So, for Durkheim, it is social institutions, be they religion or any moral order, that function in a way which brings about the integration of the individual in society and the cohesion of the social structure. This means that the social institutions impose themselves upon the individual and thus constrain him to adhere to their rules. Weber, on the other hand, stresses the importance of individual acts, the subjective meanings which individuals attach to their actions, and thus the effects they have on promoting changes in society. The different emphases in Durkheim and Weber can obviously be reconciled through a central sociological concept, namely that of the inextricable link between the individual and society. The interdependence between the two is clear: there is no society without the individual and his actions, and there is no individual without a society and its rules.

Individual Identity: Mead and Berger

George Herbert Mead (1863–1931), the pragmatic philosopher, sociologist, and social psychologist, made a major contribution to our understanding of the individual's *self*. He laid stress on two facts: (1) individuals interact by means of significant symbols, the most important of which is linguistic communication; (2) in this interaction a social process takes place in which the *self* develops. According to Herbert Blumer,

> For Mead, the self is far more than an internalization of components of social structure and culture. It is more centrally a *social process*, a process of self-interaction in which the human actor indicates to himself matters that confront him in the situation in which he acts, and organizes his actions through his interpretation of such matters. The actor engages in his social interaction with himself, according to Mead, by taking the role of others, addressing himself through these roles and responding to these approaches. (Wallace and Wolf, 1980; see also Mead, 1934 and Mead, 1982)

In the sense that we 'look at our own actions 'from the outside', as it were, we 'see ourselves', as we say, 'as others see us'. In particular, 'the significant others' in our relationships are important in building our 'self'. As Peter Worsley puts it:

> The individual actor is a different person to different Others, both because he acts differently towards them, and they to him, and because each interprets the other's behaviour differently. When the individual considers how his behaviour looks to others, he becomes a Me (the self as object of scrutiny) to his I (the self as subject, who is doing the scrutinizing).

For Mead, therefore, the individual is not 'a unitary thing called the 'person' 'since his "self" does not exist independently of others'. The individual is rather a complex entity by virtue of his interrelationship with others (Worsley, 1970).

It is when we examine an individual's action and the way it comes about that we understand best the sense in which Mead differs radically from many other social scientists who explain how the individual act is produced. Thus '[a]ction is built up in coping with the world instead of merely being released from a pre-existing psychological structure by factors playing on that structure' (Blumer, 1969). It is the individual, facing the world and sizing it up, that brings himself to act in the way he does. On this point I continue to quote Herbert Blumer at some length below, concerning how, in Mead's terms, 'the individual forges or pieces together a line of action':

> In order to act the individual has to identify what he wants, establish an objective or goal, map out a prospective line of behavior, note and interpret the actions of others, size up his situation, check himself at this or that point, figure out what to do at other points, and frequently spur himself on in the face of dragging dispositions or discouraging settings.

There may be, of course, deficiencies in the construction of the acts of any individual – deficiencies due for instance to the failure to notice things; the misinterpretation of things noticed; poor judgement; faulty mapping out of prospective lines of conduct; half-heartedness in contending with recalcitrant dispositions, and so on – but these do 'not belie the fact that his acts are still constructed by him out of what he takes into account'.

> What he takes into account are things that he indicates to himself. They cover such matters as his wants, his feelings, his goals, the actions of others, the expectations and demands of others, the rules of his group, his situation, his conceptions of himself, his recollections and his images of prospective lines of conduct. He is not in the mere recipient position of responding to such matters; he stands over and against them. He has to organize or cut out his lines of conduct on the basis of how he does handle them. (Blumer, 1969)

Mead's theory can be contrasted with that of scholars who basically explain human action by a simple formula, namely that '[g]iven factors play on the human being to produce a given type of behaviour'. Mead's scheme is fundamentally different, in that the individual is not seen as a 'mere medium for the operation of determining factors that play upon him, [rather] the human being is seen as an active organism in his own right, [and] action is seen as conduct which is constructed by the actor' (ibid.). In my opinion Mead overemphasized certain aspects of individual acts: their calculated and premeditated nature and their obvious social–interactional character. These aspects are certainly present in many of them; but one cannot disregard the importance of instinctive acts such as those produced in the circumstances of a minor accident or a major disaster, acts where no prior calculation and premeditation is possible. Neither can one neglect the effects of chance even on planned actions where the agent uses rationally information available to him in order to come to a decision, but where relevant information actually coming his way may be affected by chance. There are also cases in which making a choice is very tough because of the complexity of all the factors to be taken into account, so that people do end up sometimes making a random choice. (See Chapter 5 for a full treatment of this topic.) Mead does combine brilliantly two basic ideas: he recognizes that the human being is a 'self'; and he is aware that the self, or the person, has to face the world in an interactional social framework which includes self-reflection (Blumer 1969).

Despite the difficulty of pinning down with precision what consciousness is (a difficulty I mentioned above), Peter Berger uses this concept freely. Basing himself on Mead, he examines 'individual identity' and, in particular, the force of 'socialization' and how this process shapes identity. As he states:

> Social psychology has brought about a recognition that the sphere of psychological phenomena is continuously permeated by social forces, and more than that, is decisively shaped by the latter. Socialization means not only that the self-consciousness of the individual is constituted in a specific form by society (which Mead called the 'social genesis of the self') but that psychological reality is in an ongoing dialectical relationship with social structure. Psychological reality refers here, *not* to scientific or philosophical propositions *about* psychological phenomena, but to the manner in which the individual apprehends himself, his process of consciousness and his relations with others. (Berger 1970)

Berger goes on to explain how the individual's socialization builds up his identity. Since society contains standard identities concerning sex, age, occupation, social class, educational and cultural characteristics – to mention just a few – the individual adopts such identities according to his situation or his self-assessment, and thus he internalizes these identities. These aspects of his own identity are then galvanized in his consciousness. By simple introspection he knows who he is and how he is expected to conduct himself. As Berger maintains: 'Society not only defines but creates psychological reality. The individual *realizes* himself in society – that is, he recognizes his identity in socially defined terms and these definitions *become* reality as he lives in society.' Since individual identities are socially constructed, individual experiences can be understood in an intelligible world, which makes communication with others possible. This world is shared with others and is apprehended as 'objective reality'. Berger stresses, of course, that language is central both in the process of socialization and in the construction of reality. In summing up his analysis of how individual identity emerges and is maintained, and how human beings come to know and understand their own identity, we see that Peter Berger neutralizes psychological explanations; at best he acknowledges psychological processes as such in the context of a dialectical relationship with social forces. In the final analysis, he provides sociological and social–psychological explanations for the sense of identity the individual possesses. As he says, the sociological understanding of identity can simply be stated as follows: *'Identity, with its appropriate attachments of psychological reality, is always identity within a specific, socially constructed world. Or, as seen from the viewpoint of the individual: One identifies oneself, as one is identified by others, by being located in a common world'* (Berger, 1970).

The Individual and the Human World

In the foregoing pages of this chapter I have attempted to present as clearly as possible, and in basic terms, what is meant when we refer to an individual and his or her identity. I looked at various aspects of this identity: the genetic, the psychological, and the social. Let us now consider how the individual negotiates his relations with others and what kind of human environment he has to live in. We are not considering here the physical conditions one has to face throughout life – though it is worth mentioning that the mutual influences between the physical and the human factors are not only significant but they are crucial to an individual's life.

A useful concept when it comes to examining how one faces the world and one's own life is that of the 'lifeworld', as many philosophers and sociologists label it. In general terms, 'lifeworld' can be taken to denote the given circumstance in which an individual lives. The concept was coined by Edmund Husserl (1859–1934) in German, as *Lebenswelt*, and subsequently redefined and interpreted – more recently by Jürgen Habermas (among others). The lifeworld serves the individual as a kind of background which encompasses one's own perceptiveness, one's experiences, one's consciousness and knowledge (including knowledge of others and of their experiences), one's grasp of reality as a whole and of the various situations in it, and intersubjectivity as an integral part of one's life. Jürgen Habermas puts it as follows: 'By the everyday lifeworld is to be understood that province of reality which the wide-awake and normal adult simply takes for granted in the attitude of common sense. By this taken-for-grantedness, we designate everything which we experience as unquestionable' (Habermas, 1989).

There are a number of characteristics to the lifeworld:

1 it is intuitively present as a 'vast and incalculable web of presuppositions';
2 it is within the horizons of a lifeworld that action situations occur;
3 '[a] situation represents a segment of the lifeworld delimited in relation to a theme';
4 situations change and 'the situational horizons shift, but the limits of the lifeworld cannot be transcended';
5 any situation in the lifeworld, and any action in it, are meaningful on account of 'a social a priori built into the intersubjectivity of mutual understanding in language' (ibid.).

Regarding the two basic paradigms, that of consciousness and that of mutual understanding, Habermas seems to agree with Dieter Heinrich that 'relation-to-self and linguistic capability are equiprimordial' (Habermas, 1992).

Once we have presented, in a condensed manner, Habermas' definition of the lifeworld, his application of the concept to what goes on in reality becomes relevant to our interest in the way the individual faces the human world. One question arises immediately: since the individual confronts the human world in certain environments or situations, does this mean that his own lifeworld relates only to a segment of the 'total' lifeworld which constitutes the background for the particular situation at hand? As we shall see below, if my understanding is correct, Habermas certainly holds that such segmentation of the lifeworld takes place. At any rate, the particular situation is experienced intersubjectively – that is, *vis-à-vis* other people and *vis-à-vis* society at large. The individual is involved by means of mutual communication and understanding as to what the situation is, what the predicaments are, what action is to be taken, how that action is to be coordinated with others, how different needs and, possibly, opposing values and interests are to be solved. Habermas' solution situates itself in the perspective of 'communicative action' and within the confines of a narrowed-down lifeworld: his idea of how the individual can and ought to face the human world is encapsulated by the ensuing rational dialogue, whereby the people concerned consider each other's views, and by their consequent arrival at a democratic compromise and peaceful understanding which ultimately lead to a consensus. The communicative interaction is in fact reality in the making. Habermas holds that reality which is produced through communicative interaction is possible because rationality is located in the latter and because the goal of communication is mutual understanding.

Habermas' theory is problematic and one-sided, since he narrows down the lifeworld so as to make it complementary to communicative action. In this way he can counterpoise 'the system' to 'the lifeworld'. He sees the former as something external, intruding into the latter and thus preventing the democratic and rational communicative debate and the free dialogue between individuals, who would thus lose the possibility of arriving at peaceful agreements and, ultimately, at a consensual existence. As a result, the emancipatory aspect of communicative action is also lost. Thus, from Habermas' point of view, 'the system' exemplified by governmental bureaucracies, private corporations, stock markets, state legislatures, political parties and other regulatory bodies colonizes the lifeworld, especially the public sphere, and imposes the aims of these entities even by force, thus affecting the freedom of individuals to live their life in a way that promotes common understanding and cooperation within the groups to which they belong. It seems that such a weakening of the possibility of communicative dialogue affects adversely one's relations with the others and with society.

One cannot avoid labeling Habermas' theory as 'naïve idealism'. A critique of his theory must highlight two basic questions. First, on what grounds does Habermas assume that engaging in communicative action brings along a recourse to rationality? Second, how can the concept of lifeworld be presented as referring to the background or environment of the individual, when it excludes the systemic aspects of reality which undoubtedly affect any person throughout the whole life?

We can accept the concept of 'lifeworld' as a given, non-interpreted background of life as it is, so long as it is realistically all–inclusive; for it is against an all-inclusive background that the individual's life unfolds. Such a background includes systemic elements which are in no way external to the total lifeworld. We cannot accept some kind of complementarity, some special relationship between communicative action and the total lifeworld. The latter provides the intelligible and meaningful perspectives for the individual who has to face the human world, in a context that includes not only rational discourse, but also 'systemic and strategic exchanges of power' (Crossley and Roberts, eds, 2004). It could also be argued that, even if individual agents commit themselves to discourse, they 'can always opt out by adopting a strategic orientation'. As Joseph Heath points out, 'Habermas's argument does not show that agents *must* enter discourse, only that they *ought* to' (Heath, 2003). Similarly, irrationality, non-rational action, conflict and violence, non-resolution of problems, ambiguity, uncertainty and chance are all part of life and of the human condition. When all is said and done, the individual faces a complex web of relations which, at times, becomes hard to handle and which is often unpredictable in its ramifications and ultimate outcome. The outcome, whether fair or undesirable from the individual's point of view, depends a great deal on the period and the sort of society the individual in question lives in, the power and capabilities of others in that society, his own ability to cope with the situation he faces, and possibly his luck.

Bibliography

Adler, Alfred, 1956. *The Individual Psychology of Alfred Adler: A Systematic Presentation in Selections from His Writings*, edited by Heinz L. Ansbacher and Rowena R. Ansbacher, New York: Harper Torchbooks.

Barker, E., 1960. *Social Contract. Essays by Locke, Hume, and Rousseau*, London: Oxford University Press.

Bendix, Reinhard, 1966. *Max Weber*, London: Methuen, Chapter 3.

Berger, Peter, L., 1970. 'Identity as a Problem in the Sociology of Knowledge', in James A. Curtis and John W. Petras, eds, *The Sociology of Knowledge*, London: Gerald Duckworth, pp. 374–8.

Berger, Peter, L., and Thomas Luckman, 1966. *The Social Construction of Reality*, Garden City, NY: Doubleday.

Bettelheim, B., 1982. *Freud and Man's Soul*, New York: Knopf.

Bewaji, John, A. I., 'The Social Contract Tradition', *The Paideia Project On-Line*, p. 2, accessed 05.08.07 at http//www.bu.edu/wcp/Papers /Soci/SociBewa.htm.

Blumer, Herbert, 1969. 'Sociological Implications of the Thought of George Herbert Mead' in Walter L. Wallace, ed., *Sociological Theory*, London: Heinemann, pp. 236, 234–44.

Broom, Leonard and Philip Selznick, 1968. *Sociology – A Text with Adapted Readings*, New York: Harper and Row, 1968, pp. 26–30, 463–6 (see also Mead 1934, 94–8).

The Columbia Encyclopedia, 2001–5. 6th edn, New York, NY: Columbia University Press, *s.v.* 'Jefferson, Thomas', accessed 16.09.2007 at http://www.bartleby.com/65/je/JeffersT.html, and 'Social Contract', accessed 16.09 2007 at http://www.bartleby.com/65/so/socialco.html.

Coser, Lewis A., 1971. *Masters of Sociological Thought*, New York: Harcourt Brace Jovanovich, pp. 129–36, 227–8.

Coser, Lewis A., and Bernard Rosenberg, eds, 1976. *Sociological Theory – A Book of Readings*, London: Collier Macmillan, pp. 93–7.

Crossley, Nick and John Michael Roberts, eds, 2004. *After Habermas: New Perspectives on the Public Sphere*, Oxford: Blackwell Publishing/The Sociological Review, p. 8.

D'Agostino, Fred, 2003. 'Contemporary Approaches to the Social Contract', in *The Stanford Encyclopedia of Philosophy*, 2003, accessed 02.08.07 at http://plato.stanford.edu.entries/contractarianism-contemporary/.

Deouell, Leon, Y. et al., 2007. 'Cerebral Responses to Change in Spatial Location of Unattended Sounds', *Neuron*, Vol. 55, 2007, pp. 985–96.

de Vries, Sophia, 2006. 'Some Basic Principles of Individual Psychology', *Alfred Adler Institutes of San Francisco and Northwestern*

Washington, accessed19 Jan. 2006 at http:ourworld. compuserve. com/homepages/hstein/basic.htm (originally published in the *Individual Psychology Bulletin*, Vol. 9, 1951).

Durkheim, Emile, 1952. *Suicide: A Study in Sociology*, edited by John A. Spaulding and George Simpson, London: Routledge and Kegan Paul.

Finkelman, Paul, 2001. *Slavery and the Founders: Race and Liberty in the Age of Jefferson*, Armonk, NY: M. E. Sharpe.

Habermas, Jürgen, 1989. *The Theory of Communicative Action, II: Lifeworld and System: A Critique of Functionalist Reason*, translated by Thomas McCarthy, Boston: Beacon Press, pp. 127, 130–2.

Habermas, Jürgen, 1992. *Postmetaphysical Thinking*, translated by William Mark Hohengarten, Cambridge, MA: MIT Press, pp. 22–3.

Heath, Joseph, 2003. *Communicative Action and Rational Choice*, Cambridge, MA: MIT Press, p. 306.

Klein, Barbara, 2003. *Not All Twins Are Alike: Psychological Profiles of Kinship*, Westport, Connecticut: Praeger Publishers, Chapter 8.

Koch, Adrienne, ed., 1971. *Jefferson*, Englewood Cliffs, NJ: Prentice-Hall, pp. 19–28.

Krausz, Ernest, 2000. *The Limits of Science*, New York: Peter Lang, pp. 102–3, 130–2.

Lessnoff, Michael, H., 1986. *Social Contract*, Basingstoke: Macmillan.

McDonald, Forrest, 1976. *The Presidency of Thomas Jefferson*, Lawrence, Kansas: University Press of Kansas, p. 30.

Mead, G. H., 1934. *Mind, Self, and Society*, edited by C. W. Morris, Chicago: University of Chicago Press.

Mead, G. H., 1982. *The Individual and the Social* Self. Unpublished Essays by G. H. Mead, edited by David L. Miller, Chicago: University of Chicago Press.

Mounce, H. O., 1999. *Hume's Naturalism*, London: Routledge, pp. 90–1.

The New Encyclopaedia Britannica (Macropaedia (MEJ)), Vol. 19, Chicago: The University of Chicago Press, *s.v.* 'Freud' (Martin Evan Jay), p. 569.

Padover, Saul K., 1965. *Thomas Jefferson and the Foundations of American Freedom*, New York: Fawcett, pp. 46, 48, 51.

Popper, Karl R., 1966. *The Open Society and Its Enemies*, Vol. II, Princeton, New Jersey: Princeton University Press, Chapters 12–14.

Popper, Karl R., 1994. *Knowledge and The Body-Mind Problem*, edited by M. A. Notturno, London: Routledge.

Popper, Karl R., and John C. Eccles, 1983. *The Self and Its Brain*, London: Routledge and Kegan Paul, p. 37.

Rawls, John A., 1999. *A Theory of Justice*, Cambridge, MA: Belknap Press of Harvard University Press (original publication 1971).

Rex, John, 1961. *Key Problems of Sociological Theory*, London: Routledge and Kegan Paul, pp. 42–50, 78–9.

Ritzer, George, 1983. *Contemporary Sociological Theory*, New York: Knopf, pp. 13–6, 26–8.

Searle, John R., 1992. *The Rediscovery of the Mind*, Cambridge, MA: The MIT Press, pp. 83–4, 90–4, 116–18.

Siegel-Itzkovich, Judy, 2007. 'Hebrew University Researchers: Brain Locates Sound Automatically', *The Jerusalem Post*, 7 October 2007.

Skinner, B. F., 1953. *Science and Human Behavior*, New York: Macmillan.

Speake, Jennifer and Sarah Mitchell, eds, and Antony Flew, editorial consultant, 1979. *A Dictionary of Philosophy*, London: Macmillan, *s.v.* 'Immanuel Kant', p. 177.

'The Declaration of Independence', 2007. *Social Studies*, Glenco: McGraw–Hill, accessed 29.10.2007 at http:www.glencoe.com/sec/socialstudies/btt/celebratingfreedom/caf/shtul.

'The Mind and Brain – Synonymous or Dichotomized', 2007, accessed 26.07.07 at http:/psych-s1.psy.Vanderbilt-edu/faculty/blaker/183/Mindbrain.html.

Wallace, Ruth A., and Alison Wolf, 1980. *Contemporary Sociological* Theory, 4th edn, Englewood Cliffs, New Jersey: Prentice-Hall, pp. 224–5.

Watson, J. B., 1925. *Behaviorism*, New York: The People's Institute Publishing Co.

Weber, Max, 1930. *The Protestant Ethic and the Spirit of Capitalism*, translated by Talcott Parsons, London: Allen and Unwin.

Weymouth, Lally, ed., 1973. *Thomas Jefferson: The Man, His World, His Influence*, New York: G. P. Putnam's Sons, pp. 183, 189, 229.

Worsley, Peter, 1970. *Introducing Sociology*, Harmondsworth: Penguin Books, p. 47.

5 The Elements of Rationality and Chance in the Choice of Human Action

I shall concentrate in this chapter on deliberate human action: action that is preceded by propositional attitudes, which Bertrand Russell regarded as the beliefs, desires or doubts that human beings constantly apply when deciding on an action (Russell, 1940). It is to propositions, arising out of linguistic communication, that attitudes are taken by human beings (Donagan, 1987), becoming the intentional attitudes which are 'directed upon a content . . . and in that sense an object' (Audi, 1993). In other words, the attitudes are transformed into intentions and then into actions, which take place once human beings make their choices for action. For my purposes there is no need to summarize the elaborate explanation of what 'intentionality' is, as this topic has been fully treated by John Searle. It will suffice here to draw attention to a major distinction he makes between 'prior intention' and 'intention in action', which causes the action to take place and thus achieves a preset goal. This leads to the further distinction between:

(a) planned action, which involves both prior intention and intention in action, such as when I want to invest in the stock market or I think of undertaking a journey;
(b) spontaneous action, which is not preceded by prior intention, it is not planned, but it does involve intention and planning in

action – otherwise the action could not be carried out – as for example when a sudden obstacle on the pavement forces me to take spontaneously an avoiding action, which requires intention and instant planning; and

(c) casual unthinking activity, which does not require intentions, as in the case of fidgeting or yawning during a dull lecture (Searle, 1983; Wakefield and Dreyfus, 1991).

Case (c) is of no interest to me in this context. Case (b), although it involves both intention and deliberate planning, even if of a very short duration, is not as useful for me in the present discussion as case (a) is. The latter provides the kind of terrain which is conducive to sorting out the elements involved at the intentional stage – that is to say, the stage at which the agent deliberates about possibilities, given his goal, and he chooses his action in order to achieve it (Donagan, 1987). It is at this stage that propositional attitudes are applied, that rationality enters and that chance possibly plays a role. The central questions I wish to address are whether purely rational choice is possible and whether choices may be induced by chance alone, or whether there is always a mixture of rationality and chance, as well as of other factors such as habit, imitation and irrationality, in the choice of an action.

I should predicate my inquiry into 'choice of action' by stating that the discussion pertains to action which can be explained by its antecedents such as the reasons and preferences of the agent to act in a certain way; the particular state of mind in which the agent is when deliberating what action to choose; and the state of the world in which he chooses and carries out his action. All these factors influence his choice, yet this is not incompatible with the notion of 'free choice' – in other words, his action is not pre-determined (Ginet, 1997).

Let me elaborate somewhat on the above statement. The question is whether the very fact that the agent is weighing up his options, in a quest to decide what action to take, will or will not alter the factors which we call 'the antecedents'. Facing the need to decide what action to go for may make him shift his preferences, change his attitudes and frame of mind, and 'see' the state of the world from the particular vantage point of the choice situation. It might still be claimed that, at time *t'* when the choice, say, as to how to invest in a volatile stock market or how to travel during a transit strike was made, there were certain antecedents X, Y, Z, which determined his choice. My argument, however, is that it is a fact that at time *-t'* when he became aware of the need to make a choice, his preferences, his state of mind and his view of the state of the world might have been different from what they became at time *t'*, due to the flexible nature of these factors during the period of deliberation. My claim is, therefore, that an action is not pre-determined by some previous, rigidly established, factors, but that it is an outcome of the process of choice itself. My approach is that of a factualist; hence I do not take into account the metaphysical tradition which upholds the deterministic view that, since all the factors leading to and producing the choice of action were antecedent to it, the choice was in that sense pre-determined. Such a view would acknowledge the existence of a mind-independent world, as realists propose (Khlentzos, 2001). Thus, although the realists would not deny the apparent possibility of human beings choosing among alternative actions, according to them the actual choice will have been pre-determined. A factualist approach, on the other hand, would uphold the view that it is the actual choosing process that determines the final choice of action. This complex process includes the sifting and evaluation of the received information and the ensuing deliberation, based as the latter might be on rational considerations, but also probably affected by existing habits, doubts, and the random selection and non-evaluation of some of the possibly relevant information. Whatever the sources of the elements involved in the choosing process, the choice of action is a specific outcome, created by the acting agent. Various outcomes are possible, even given the same baseline sources and conditions, due to the different possible interactions, including chance interactions, between the various factors during the process leading up to the actual choice and action.

Choice, Rationality and Chance

Before dealing with the major questions posed, it will be useful to clarify the above concepts, as well as some ideas connected with these concepts.[1]

Choice

Choice can be defined as the decision to act (or not act) in a certain way when one is faced with one or more possibilities of action. Making a choice appears to be the initiating part of a process whereby the actualization of an action or the realization of inaction takes place. Choice is part of the world of propensities described by Karl Popper. He holds that 'propensities should not be regarded as properties inherent in an object, such as a die or a penny, but they should be regarded as inherent in a situation (of which, of course, the object was a part)' (Popper, 1995). Similarly, the expression of propensities by means of 'choice' can be seen to be external to the objects or social acts available to choose from. 'Choice' is thus part of a situation that provides the opportunity for action and necessitates a decision. As stated in the definition above, deciding not to choose from an array of possible actions is itself a 'choice' and it, too, produces an effect. This point is made by Searle when he says that, '[c]haracteristically, human intentions are designed to produce some further effects on the world' (Searle, 1991). It is of no concern to us whether the actual choice is determined by past events or by expectations concerning events in the future; whether such past events or expectations give rise to motivations to choose one way or another; or whether such choice can be spontaneous, that is to say, capricious, rather than weighed rationally (see below on the issue of rationality and choice). But, whatever its roots, it is certainly the case that a choice means the beginning of actualization, or realization in the case of inaction – unless obstacles arise after the choice which prevent such actualization. The actual carrying out of the choice not only crystallizes the situation but momentarily freezes it. However, the choice will often have wider and longer-term effects, since it will cumulate and intermesh with other choices, of individuals or collective bodies, to produce a new situation, with new possibilities and propensities (Popper, 1982; Sfendoni-Mentzou, 1993).

It must be stressed that choices do not have an unlimited scope; they are always made within certain constraints. A poor man cannot choose to live in an expensive villa, although he can strive to achieve the wealth necessary to do so – a striving which, irrespective of the outcome, will lead to various choices on the way.

Another type of constraint is that, once a choice is made, in a sense it cannot be unmade; it will have an effect that cannot be nullified, although new and contrary choices can alter the direction of behaviour that the earlier choice signaled. It is only before the choice is implemented that it can be reconsidered and another choice adopted. Very often, but certainly not always, when the decision-maker makes his choice, he will take into account the conceivable consequences of his action, which ensue. He may or may not be aware of the whole sequence and all the ramifications of the choices he will have to make following the choice he has already made (Levi, 1997). Next, choices have outcomes not only for the chooser but also for others; consequently choices are integral parts of social contexts. Finally, whilst choice signifies a certain openness since it has to do with the future, once an actual choice among various possibilities is made it becomes exclusive. Yet, although it is true that in the existing situation it cancels out other choices, the very choice adopted changes the situation and thus opens up new possibilities and new propensities (Popper, 1995).

Rationality

It is difficult to provide a precise definition for 'rationality', but human linguistic communication and the ability to classify instances, as well as having notions of truth and falsity, are regarded as the foundations of all our reasonings (Balzer, 1993) and as giving human beings the ability to think rationally. In a popular sense and in dictionary definitions, 'rationality' is counterpoised to religious beliefs or supernatural revelation, being regarded as the capacity to 'reason' and thus to accept explanations of phenomena only when they are tested in a reasonable way by actual experience. With Descartes' emphasis on the existence of the thinking self and acceptance of a world open to rational exploration, humanity entered 'the enlightenment' or the age of reason, where rationality is proffered as the best way for attaining knowledge by the thinking conscious human being.

We do not need to delve into the different approaches of Hume and Kant, one stressing the sentient human being as the database or starting point of rational thought, the other stressing the capacity of the human mind for *a priori* reasoning. Whatever the approach, reason was seen as something universalistic and built into our thoughts – a kind of inner logical compulsion (Gellner, 1992).

J. H. Plumb observes that 'man's rationality, his capacity to think about himself and his environment and to communicate his thoughts is obviously his most precious gift and makes him unique in the animal world' (quoted in Gould, 1971). Julius Gould sums up and defines what we mean by rationality when he calls it 'a shorthand way of pointing out . . . the canons or standards which I and others appeal to when we try, individually or collectively, to give our "reasons" in a clear and demonstrable way'. Elaborating on the concept of 'rationality' in conduct and belief, we find that it means many things: the ability to reflect, to criticize and argue logically; to act in a manner appropriate and relevant to a situation; to be consistent by acting in accordance with one's purpose; and at the same time to take account of the purposes of others (Gould, 1971). In practice, whether one is deciding on what goals to seek or how to achieve them, rationality is expressed by responding to one's own presentation of reasons, as well as by taking note of, or seeking, the input of others. A consequence of such a multi-faceted capacity is one's ability to choose among alternatives and to scale one's preferences (see the assumptions of 'rational choice theory' below).

The ability to think and act rationally has, of course, been hailed by many as the progenitor of our 'age of science' (Newton, 1997). But, although human beings have the capacity for rational choice and rational action, they do not always act or choose in a rational way. In cases where they do not, we label their choice/action as irrational. Can the distinction rational/irrational be based on some objective criterion or measure? One such criterion might be the adequacy or appropriateness of an action given a particular situation, as in the case of a war situation. Assuming that a war leader's purpose is to win the war, his action may seem, from a wider historical perspective, to have been inappropriate and thus 'irrational'; although, considering the way he saw the situation at the time, his action might have seemed appropriate, that is, 'rational'. As Robert Audi puts it: 'One might think, then, that the crucial mark of a rational action is its appropriateness to the aim(s) of the agent at the time of action' (Audi, 1993). But this simply means that we can understand why the agent took the action, which in the wider view appears to have been inappropriate (Popper, 1994) and, therefore, irrational. However, rationalism claims that all human action has a rational basis. This becomes apparent if we adopt multicultural sensitivity and we dig deep enough to find the rationale of an action. In other words, within a particular culture or from a specific viewpoint of the agent, every action has its own logic and rational basis. Taken to the extreme, this claim means that what appears irrational from a western cultural standpoint, or from the point of view of the standard adopted by a particular observer of an act, is not in fact irrational. That is to say, all actions have their reasons and inner logic; 'all actions to be actions must be rational at some level' (Fay, 1996). Brian Fay and other critics reject such a conclusion, insisting on the possibility of irrational choice and action.

It is the instrumentalist conception of rational action that reinforces the idea that basically all action is grounded in rationality. As John Searle points out, the classical model of rationality is based on a paradigm which emerged from Wolfgang Kohler's experiments (early in the twentieth century), which showed that apes engage in rational action by using the necessary means (box and stick) to achieve their desired ends (take down a banana from a high ceiling). The classical model, therefore, 'represents human rationality as a more complex version of ape rationality'. Searle is critical of the classical model. In his view, '[i]n human rationality, as opposed to ape rationality, there is a distinction between reasons for action which are entirely matters of satisfying some desire or other and reasons which are desire-independent' – as in the case where I do not desire but am obliged to take an action. Finally, although Searle emphasizes that 'constraints of rationality are universal and built into the structure of mind and language . . . and hence all argument presupposes rationality', yet in his view, inevitably, 'you can only have rationality where you have the possibility of irrationality' (Searle, 2001).

Having reviewed briefly the notion of rationality, I adopt the views of John Searle and Brian Fay that, although rationality is a most basic faculty of human thought, not all human action is rational. Thus 'agents may have reasons for their actions, even though these reasons may be irrational' (Fay, 1996).

Chance

First, what exactly is meant by 'chance', which Ernest Nagel labels as 'a notoriously ambiguous word'? There are various senses in which the word can be used, for instance to describe the occurrence of an unexpected event, to trace an event to the fortuitous intersection of independent causal chains, or to denote contingency –that is, the property of not having to exist or to occur (Nagel, 1961). In none of these cases is it denied that, when an event labeled as a 'chance event' does take place, that event is caused. The basic question is whether we invoke Poincaré's notion of 'chance', which stipulates that chance is only a measure of our ignorance: that is, 'we ascribe an event to chance not because it has no cause but because we simply do not know the cause' (Poincaré, 1905; see also Hwang, 1993). On this view, ultimately the deterministic position can be defended. Alternatively, if we accept the idea of absolute chance, which refers to a spontaneous uncaused event, then the extreme indeterministic position will have to be taken up.

The notion of 'chance' and its possible meanings have been extensively debated, from Aristotle to Peirce and to more recent philosophers. Much of the debate turns on whether Aristotle's *tuche* (chance) can also mean pure chance – that is, whether it implies that events can happen without any cause whatsoever, being self-determined; or whether the opposite contention is true – namely that chance-facilitated effects do have causes, but are characterized by freedom and spontaneity, or by lack of intention from an acting agent, as in the case of an accident (Hwang, 1993; for a discussion of causation and chance, see Krausz, 2000).

As an area of human endeavour in which chance enters a great deal, it is worth mentioning here the area of scientific research. There are many examples of chance occurrences and serendipity in scientific discoveries (Krausz, 2000). As R. Taton points out,

> A certain number of examples of accidental observations leading to fundamental discoveries has given credence to the opinion that chance plays a primary role in scientific creation, and that many discoveries would not have been made without the fortunate occurrences of very exceptional circumstances, without which any efforts

> on the part of the research workers would have remained unfruitful or at least unrewarding. (Taton, 1957)

At this stage it will suffice to sum up the meanings of chance as follows:

(a) chance reflects our inability to predict single events, such as the number that will turn up at the throw of a die;
(b) chance is associated with indeterminism, randomness, irregularity, change, fluctuation and instability, exemplified by error rates in cell reproductions and spontaneous mutations, or with the existence of random events represented by radioactive decay, or by the numerous fortuitous occurrences in human relations and in the processes of human cognition;
(c) chance is associated in the social world with human will, that is to say, with the ability to choose between alternative actions;
(d) chance represents the idea of potentiality and actualization, namely that only some of the possible outcomes or events will become actualized in reality.

Finally, to give my own view of chance, I would say that it represents the accidental combination and cross-cutting of factors or circumstances which result in unanticipated events. Neither such occurrences nor their ensuing results are predictable. Furthermore, such occurrences are often unbeknown to agents and do affect their presumed rational choice and decision-making. In other words, the very alternatives available for choice and the circumstances in which the choice is made are influenced by chance events which are unpredictable and uncontrollable. So it is chance as defined in (b) that is of particular interest to me, in so far as it affects human action. Let me also make it clear that my interest is not in how chance can be emprically assessed, as it is by laws of statistical probability (Mellor, 1971; Burks, 1977), but rather to show how chance enters the reality of the human choice of action.

The Process of Choosing an Action

There is a whole range of situations in which individuals, or groups of people, will be faced with the need to choose from a number of possible alternatives and thus decide upon a course of action. In some situations the options are clearer than in others. Let us assume, however, that the options are absolutely clear, in which case we can set up axiomatic models prescribing how the decision-maker should act if only rational calculations are made. One such example is the model based on game theory, which assumes that the players wish to maximize their profit or some other advantage, or wish to minimize the possible damage, and will do so by means of rational calculation following certain logical steps, given the situation they are in. The situation is competitive, and the outcome depends not only on one's own choice but also on that of others involved in the game. Therefore there is an attempt to anticipate the probable choices of others in order to make one's choice 'the best choice'. Such models assume, as I said, that those involved will make rational calculations concerning how best to achieve their goals, the calculations depending on their ability to rank-order the possible outcomes by attaching utilities or values to them. I do not deny that in real situations, whether in economic competition or in war strategies, people will attempt to make rational calculations concerning the best way of achieving their goals and try to predict the outcomes of their choices. However, it is also undeniable that players in a real situation may have attitudes, needs and expectations which lead them away from the rational choices that the game theory would prescribe. Thus even the costs and expectations involved in playing the rational game may lead the individual to choices based on tradition, habit, emotion, imitation, and so on, rather than on rational calculation. Where material benefits are concerned, physical costs (tiredness), psychological costs (the concentration needed, instead of attention to other interests), time costs (time not available for other actions with possible material benefits), or expectations (prejudices concerning the abilities of others) – could all lead to other than immediate rational calculation in making a decision. It is true, however, that, if such weighing up of costs and expectations does go on in the mind of the individual, this in itself points to rationality in his conscious thinking.

In some cases of human behaviour, where the aim is purely to achieve material benefits and satisfaction, one could argue that it is always rational behaviour that determines the choice of action.

Thus 'rational choice theory' is used in economic analysis, being based on the assumptions that the person knows what he wants, orders his wants according to a scale of preferences and chooses his action rationally, so as to maximize his satisfaction. As James Rule points out, it cannot be denied that at least two basic tenets of rational choice theory are in evidence when we consider practical decision-making: one, that 'people's actions are governed by relatively enduring hierarchies of preferences', and two, that 'people weigh different courses of action, altering their behavior when changed strategies give better promise of producing the desired results' (Rule, 1997).

There are, however, many reservations concerning the rational choice paradigm. Some telling comments about it and its shortcomings come from an expert in game theory modeling:

> The basic motivation for studying models of bounded rationality springs from our dissatisfaction with the models that adhere to the 'perfect rational man' paradigm. This dissatisfaction results from the strong tension arising from a comparison of the assumptions made by economic modelers about 'perfect rationality' with observations about human behavior. (Rubinstein, 1998)

Rubinstein adds a long list of caveats concerning rational choice: (1) 'decision makers tend to simplify choice problems, probably as a method of saving deliberation resources'; (2) 'agents . . . systematically do not make correct inferences'; (3) the possibility exists of wrong timing, where a decision is taken before relevant information is revealed; (4) there are constraints and costs involved in acquiring the information necessary for the decision-maker; (5) '[i]mperfect recall is a particular case of imperfection of knowledge that lies at the heart of human processing of information'; and (6) there is the factor of the complexity of group decisions. Rubinstein refers to the experimental work of Tversky and Kahneman (1988; see below) and remarks: 'We now have a fascinating compilation of experimental data demonstrating the circumstances under which rationality breaks down and other patterns of behavior emerge' (Rubinstein, 1998).

The need to work with a new concept was highlighted many years ago, when the idea of 'bounded rationality' was introduced by Herbert Simon in the mid-1950s. Simon criticized the value of utility maximization as a normative choice principle and 'proposed the alternative normative principle of satisficing'. As Ray Crozier and Rob Ranyard put it, Simon 'argued that people can successfully adapt to their environment by identifying actions that are merely satisfactory or suffice, rather than a potential maximum'. This principle 'could be applied without sophisticated powers of discrimination and evaluation, powers that humans do not possess'. Consequently Simon 'developed the concept of bounded rationality, which incorporates the basic assumption that rationality is relative to the information processing capacities of the decision maker' (Crozier and Ranyard, 1997). Bruce Kaufman's comments are of interest here. He mentions that Herbert Simon drew part of his inspiration for the concept of bounded rationality from the work of John Commons, and adds that 'Commons described the typical human condition to be one of purposeful behavior, mixed with significant elements of "stupidity, passion and ignorance"'. Kaufman dubs these characteristics by the more elegant descriptions of 'limited mental cognitive ability' and 'limited knowledge' – these two limitations having been adduced by Herbert Simon for his proposal of 'bounded rationality', while he ignored the third factor of 'human emotional states'. Kaufman adds:

> This is a serious omission . . . for while emotions are in many ways a functional adjunct to behavior . . . extremes in emotion due to feelings of intense anger, frustration, depression, love, jealousy, lust and so on can seriously or even completely disrupt logical, rational thought processes and thus create a condition of bounded rationality. (Kaufman, 1999)

The choice/decision process is certainly a complex one. Not only has the decision-maker to decide what is desirable, what is feasible and what will be the consequences of his chosen action (in his estimation), but he also has to consider how and when to decide. Furthermore, 'decision makers also have to make inferences about the environment in which they operate, an activity dependent on their ability to analyze the situation' (Rubinstein, 1998). The intricacies and uncertainties in the decision-making process entered into by an individual can be further illustrated. Thus Hilary Putnam draws our attention to the complex nature of preferences and to the illogicality and incoherence that may appear in the sincere verbal reports of such preferences, as well as to the difficulties of interpretation (Putnam, 1983). Furthermore, Amos Tversky and Daniel Kahneman have demonstrated that people 'systematically violate the requirements of consistency and coherence', which are elementary in rational choice. And, most importantly, they point out that there are alternative frames for a decision problem and that the evaluation of options differs according to the frame a decision-maker adopts, the latter being 'controlled partly by the formulation of the problem and partly by the norms, habits and personal characteristics of the decision-maker'. The framing of the acts that an individual can take, the contingencies involved and the possible outcomes – all influence choice. The fact that the relative attractiveness of options varies according to the framing of the same decision-problem in different ways does not necessarily imply irrationality in people's choices, since people usually act 'on the most readily available frame', generally being unaware of or unwilling to explore other possible framing for decision-making problems (Tversky and Kahneman, 1981). But certainly it is clear that, facing a decision problem, there is no one clear rational choice which will always be made given the options, but that there are a number of possibilities, depending on the frame into which individuals put their decision problems. Therefore the limitations of the rational choice paradigm cannot be doubted.

There is also no denying that people's choices and actions vary in the 'degree of rationality' – a still serviceable concept of Max Weber – from the highest degree, where both goals and means are chosen rationally, to actions based on purely spontaneous emotions, where there is no rationality at all (Weber, 1964). But this kind of traditional view – that emotions can be arational (devoid of any rationality) and that they are contradictory to rationality and pose obstacles to it – has been challenged more recently, especially in cognitive psychology and neuroscience. The view gaining ground is that emotions and rational thinking are closely intertwined (Evans and Cruse, 2004). Thus Antonio Damasio regards emotions as a necessary part of rational thought (Damasio, 1995; Damasio, 1999). Similarly, Robert Solomon maintains that 'emotions are rational and purposive rather than irrational and disruptive'. But he admits that they

> differ from 'cool' judgments and normal rational deliberate action, in that they are prompted in urgency and in contexts in which . . . considered judgments will not suffice. An emotion is a necessarily hasty judgment in response to a difficult situation . . . It is the situation, not the emotion which is disruptive and 'irrational'. (Solomon in Calhoun and Solomon, eds, 1984)

According to Solomon, emotions are rational in that they are based on judgements and intuitive logic, their aim being to work out strategies for the maximization of self-interest – although, once again, Solomon can see an irrational aspect in emotions, in that often 'their strategies are hopeless or just plain stupid' (Solomon, 2003). Another positive approach to emotions is that of Martha Nussbaum, who views emotions as 'intelligent responses to the perception of value', and holds that '[e]motions are not just the fuel that powers the psychological mechanism of a reasoning creature, they are parts, highly complex and messy parts, of this creature's reasoning itself' (Nussbaum, 2001). However, others have criticized the view that emotions are inherently rational. It is precisely because emotions are intrinsically linked to value structures, which are often maladaptive, that emotions and behaviour can be typically irrational (Sripada and Stich, 2002).

Let me say at this point that my concern is not with the question whether emotions or moods are inherently rational, but rather with the proneness of such feelings to chance. Emotions are evoked by both physiological and mental states; they are the sudden responses to changing events; and they are to be understood as 'upheavals of thought' (Nussbaum, 2001). These are ingredients which provide a fertile ground for the play of chance in the thinking process, which includes the specific task of making choices (see my reference below to the thinking process and chance).

Here we may also bring in the notion of ambivalence, which, according to Neil Smelser (1998), 'leads us to understand and explain a range of behavior and situations beyond the scope of rational choice explanations, however far the latter may be stretched'. Ambivalence represents the psychological state of an individual in which he simultaneously holds positive and negative feelings, attitudes or orientations towards people, objects or symbols. It is thus altogether different from mental states where rationality rules: (a) the latter means that individuals have univalent orientations, that is, clear-cut preferences; (b) preferences are regarded as relatively stable; (c) rationality is at play in a state of consciousness. Ambivalence, on the other hand, involves uncertainty in orientation, lacks stability, and its dynamics and change 'occur beyond the range of consciousness and calculation' (Smelser, 1998).

To gather the strands in this section on the process of choice, I wish to distinguish between different choice situations and to stress the variability in their duration and flexibility. Thus, in the short-term choice situation, there are such constraints that, once the choice is made, it is irrevocable. I cannot change my choice to travel by car to my work-place when I am trapped in a huge traffic jam. There are, on the other hand, long-term choices where changes can still be made: this is true of the chain of decisions a military commander will have to make when conducting a battle, or of the decisions a sales director has to make, whose aim is to reach a certain volume of sales. The latter type of choice situation leads Lee Roy Beach to stress the point that in many choice situations there is a great deal of flexibility. Very often the decision-maker has no clear idea of a potential set of options, and his process of choosing does not start with a clear notion of what might be reasonable plans of action.

> Often you start off in one direction only to change your mind when things go awry. Rather, the process seems to feel its way along, changing in the light of feedback and often leading in directions that never were conceived of when it all began . . . The 'feeling along' nature of decision making is very difficult to capture. . . . It is far easier . . . to talk as though decision options actually exist in full-blown clarity and that the choice among them occurs at a single point of time. But this is merely a useful fiction. (Beach, 1997)

In concluding this section on the process of choosing an action, it seems to me that the possibilities for rational decision-making in choice situations are uneven: they range from clear options, a maximum capacity for analysing the situation, and a clear striving on behalf of the decision-maker to rely on, or favour, rational thinking, thus allowing a very high degree of rationality in the choice made; to unclear options where the ability and desire of the decision-maker to analyse the situation in rational terms is at the lower end of the scale, resulting in irrational or chance decisions. An interesting choice situation is one where a person is indifferent to two possibilities and chooses between them by tossing a coin. In such a case, the decision to leave the choice to chance is itself rational, but the actual choice is determined by chance.

The Role of Chance in the Choice of Action

The last example I have given is of the kind of 'chance' that can be brought under the rules of statistical probability, which John Venn (1962) distinguished from 'chance' in the form of 'mere coincidences', which are 'simply cases in point of independent events . . . which we cannot bring under any special generalization: certain, probable, or even plausible'. The rules of statistical probability apply to large numbers, such as the succession of heads and tails in tossing a coin many many times, or in the distribution of male and female births over a period in a population. It is quite a different matter when a rare occurrence takes place, such as an unplanned meeting with a long lost cousin from Australia after forty years, in exactly the same place and on the same day of the year as it was the case when the cousins said good-bye to one another. Venn called such an occurrence a 'casual coincidence'. Jacques Monod similarly distinguishes between phenomena where the concept of chance and the theory of probability are used, so that uncertainty is purely operational, as against cases where chance takes on an essential meaning of uncertainty. He calls the latter 'absolute coincidences', 'which result from the intersection of two totally independent chains of events'. His illustration is as follows:

> Suppose that Dr Brown sets out on an emergency call to a new patient. In the meantime Jones the carpenter has started work on repairs to a roof in a nearby building. As Dr Brown walks past the building, Jones inadvertently drops his hammer whose (determin

> istic) trajectory happens to intercept that of the physician, who dies of a fractured skull. Chance is obviously the essential factor here. (Monod, 1972)

Both types of 'chance' are of interest in so far as choice of action is concerned. Let us take another example: that of a young couple, Jane and William, who met at a party and eventually decided to get married. Both cognitive and affective elements will have entered their behaviour and the thinking process which led to the decision. The rational thinking process does not exclude emotions and desires; neither are emotions and desires devoid of rationality (Baron, 1994; Glanz, 1989). Such decisions by young couples are, however, 'chance loaded', since Jane met William rather than George at the party. This is a 'one time' chance event; it is not a chance occurrence that will be repeated many times. According to Venn's distinction or Monod's definition, it falls in the category of a 'coincidental chance event' (Venn, 1962; Monod, 1972). But other chance events, which do come under the rules of probability, were also part of the decision-making process. Thus, basing themselves on certain emotional states such as being attracted to each other, or on material gains expected should they get married, Jane and William were able to arrive rationally at a positive decision, which was only one in a number of possible results of their calculations. The particular mood they were in, the kind of material gains to be made, status considerations and so on could vary and thus be given to chance. The actual outcome, as one in a number of possible outcomes, is governed by the rules of probability; but it is obvious that chance played a role in the direction taken by Jane and William during the thinking process, which led to their particular choice/decision. It is clear that in this illustration both absolute coincidental chance and chance events which come under the rules of probability played their parts in the actual outcome. Since there were reasons behind the decision taken by Jane and William, I shall refer here, briefly, to Roy Bhaskar's theory of 'agency'. He holds that 'intentional human behaviour is caused, and that it is always caused by reasons', and that any such intentional behaviour is preceded by social forms, which are a necessary condition for the agent's action. 'The causal power of social forms is mediated through human agency', and thus 'the causal power of human agency' is established (Bhaskar, 1989). In a nutshell, the outcomes of agency are caused by reasons. But, surely, one can argue that the outcomes are also caused by chance, not just by reasons. This is obvious in the case of 'absolute chance', such as in the case of Dr Brown being struck by Jones' hammer. But it is also true of chance which can be subsumed under the rules of probability. It is 'reasons' that come into the rational calculations of Jane and William leading to their decision; but, since it is a matter of chance which way these calculations will go – they could have ended up in either Jane or William, or both, deciding not to get married – the outcome, in that sense, is caused not only by reasons but also by chance. This actually highlights Bhaskar's view of the openness of the social world, a world of open systems 'where invariant empirical regularities do not obtain'. The outcome brought about by the agent's choice, where deliberate intention and rational calculations are involved, is nevertheless one of contingent causality, since chance is an integral part of the entire causal structure.

It is obvious that I now have to make reference, albeit briefly, to the concept of 'thinking', since I wish to show how the choice of action is affected by the 'chance' factor – not only insofar as external events affecting choice are concerned, but also with regard to the effect of 'chance' on the very cognitive process of thinking and decision-making. The term 'thinking' has lost its appeal in current psychological parlance, despite its usefulness. It is true, though, that we do come across the difficulty of defining what 'thinking' is. Some regard it as 'the manipulation of information' (Radford and Burton, 1974). Frederic Bartlett regarded 'thinking' as 'the extension of evidence, so as to fill up gaps by moving through a succession of interconnected steps' (Bartlett, 1958). Behaviourists define 'thinking' as simply subvocal or silent speech, that is, as nothing but talking to ourselves – which is not a particularly helpful definition (Radford and Burton, 1974). Let me adduce two more views on 'thinking': one is that in the process of thinking 'we are somehow referring to, in cognitive touch with, what we do not at the moment see or feel', in other words that it is 'cognition in absence' (Price, 1969); another view is that 'thinking refers to mental experience or the contents of consciousness' and is concerned with 'the mediation of activity' (Johnson, 1972). To sum it up: by the process of thinking we are able to use consciously evidence or to manipulate information concerning options of action 'in which the possibilities are strengthened or weakened on the basis of evidence, in the light of the goals'. But in the real world, as Jonathan Baron says, 'goals are often complex and it is difficult to "purify" the situation so as to determine how a subject would pursue a single goal'. Yet in Baron's view rationality is 'the kind of thinking that helps us achieve our goals', although he accepts that 'rationality is a matter of degree' (Baron, 1994).

The possibility of chance entering the thinking process can be shown through the many experimental studies which point to changing mental states where 'chance' is bound to make its mark. One cannot ignore a strong point, made by Aaron Sloman that, in the case of the mental process, which exhibits such a very complex system, 'inferring inner states from behaviour with any reliability is nearly impossible' (Sloman, 1993). Yet we find a good deal of psychological literature, citing experimental findings, which highlights the imperfections of reasoning by human beings – areas where chance is most likely to be at work. A few examples will suffice: cognitive dissonance (Festinger, 1957; Wicklund and Brehm, 1976; Plous, 1993); memory bias (Loftus, 1980; Myers, 1990); context dependence (Coren and Miller, 1974; Milgram, 1974); mood dependent memory and evaluation (Worth and Mackie, 1987; Martin, 2001); the influence of feelings on judgement (Martin and Clore, eds, 2001); and the influence of framing (see above). In addition to such imperfections occurring in the process of thinking and reasoning and making it prone to the effects of chance, I would ask the following question: what if, in the process of analysing the information concerning options which are open to the potential acting agent, a sudden stream of thought, such as remembering some other essential goal, deflects the first stream of thought, which attempts to arrive at a reasoned decision, and as a consequence of this conflict the agent decides to take the 'best guess'? The 'memory jogging' which demanded immediate attention to some other goal might have been triggered by a chance association of ideas or by some experienced external event, quite by chance.

There is, of course, a whole host of 'chance' factors which can affect the process of rational thinking and decision-making. The agent may on some occasion not absorb or understand fully the information coming his way. He may, by chance, be aware or totally unaware of information relevant to the decision to buy an item, depending, for instance, on whether he was able to view television advertisements or read 'consumers' surveys and reports'. There may be memory errors or an inability to recall relevant information. Interference by members of the family or by colleagues may affect one's rational thinking. Fatigue and lack of time provide more examples of what may enter as chance factors and affect one's decision. Can such factors be regarded as 'mere noise', affecting only marginally the reasoned weighing up of available options? Certainly not. Many of the situations are chance related and can seriously affect choice decisions. But where the cognitive limits of the decision maker lead to non-optimal choices, these are explicable and predictable, and are not due to chance.

Why has 'chance' been given so little attention by social scientists when considering the thinking process in decision-making? I do not refer to the voluminous literature concerning possible outcomes – that is, chance as reflected in probabilities. My criticism concerns the unwillingness to give up the basic assumption of rational behaviour in situations of choice. So, despite the possibility that the agent's choice is constrained by his lack of knowledge of which state of nature, out of a number of possible states of nature, is the true one, and of which consequence of his action, out of a number of possible consequences, is likely to occur, economists have continued to resort to, and seek, a solution by declaring such a situation as 'choice under uncertainty', to which probability theory can be applied. Namely, the agent remains a rational individual, and it is held that 'any individual will act in a manner which may be described as saying that he assigns probabilities to the different states of nature and then chooses his action among those available so as to maximize expected utility' (Arrow, 1984). As Kaufman says, 'in nearly all neoclassical models the core assumption of rationality is maintained – that the human agent has the mental ability and motivation to calculate optimal outcomes' (Kaufman, 1999). In this way the probability assessments by agents control their actual choice, and the possible effects of 'chance' are thus ignored. Can one not imagine a situation in which a decision-maker will not attempt to calculate the probabilities of the state of nature or the consequences of his action? He may choose quite by chance, spontaneously, without weighing up what might be the 'best choice' for him, without reference to the states of nature or to the consequences of his action. He may also choose under the influence of a dominant 'role model', or a highly charged group event in which he takes part, without weighing up the pros and cons of his choice and consequent action.

Another reason for the neglect of 'chance' is the social scientists' preoccupation, especially in cognitive psychology, with experimental research in which laboratory methods are applied, which are detached from decision-making in natural settings – an example being the research on 'gambling' as a task for studying decision-making. As Crozier and Ranyard say:

> the gambles [in the research situation] are typically stripped of many of the features of 'real-life' gambling and it is arguable that the conditions of presentation serve to exclude many of the psychological responses that characterize risk taking . . . Rather than a research strategy that assimilates preference and uncertainties to a paradigm of rational behaviour, it might be more fruitful to describe them as they function in real-life decisions. (Crozier and Ranyard, 1997)

The focus on real-life situations means that we have to deal with decision situations which are complex and with deliberations by the agent which are most complicated (Levi, 1997); and, above all, we must allow for the possibility of 'chance'-determined choice of action. The complexity of deliberation by the agent is exemplified by our lack of knowledge of how, in our thinking, conscious processes interact with unconscious processes (Gigerenzer et al., 1990), and how exactly intuition works in decision-making – intuition dealing with and processing 'a wide array of information on many levels and [giving] us an instantaneous cue' (Agor, 1986). It seems to me that we have to revise basically our approach to how the choice of action is decided upon. I am aware of the possibility that decisions influenced by chance and intuition may be regarded as spontaneous and non-contemplative. These kinds of decision, it may be argued, do not fit with Shackle's definition of 'decision' as

> [a] transition between a psychic state in which the individual is holding in contemplation a number of distinct action-schemes each of which is open to him, to a psychic state in which he has mentally committed himself to one of these schemes. (Shackle, 1958, quoted in Levi, 1986)

It may well be that such a contemplative process necessitates the analysis of information and the rational weighing up of options in an ordered thinking stage, before any commitment to action takes place. However, as I argued above, in thinking itself, the possibility of 'chance' entering the process is always alive – its existence cannot be denied.

The Limits of Rationality

As seen from the foregoing, the aim of my paper has been to show how rationality in human action is circumscribed by the effects of chance. However, before my concluding arguments, I wish to temper this claim from two perspectives:

First, I do not deny the crucial role of reasoning, including rational calculation, in human agency. More widely put, human agency is truly an intervention in the flow of events and is part of the causal structure of human life. Human actions are not simply happenings by chance. Human beings have the capacity of reasoning and of a higher order reflection that we call rationality, when they decide to act and they choose among different alternatives (Velleman, 2000; Bhaskar, 1989). It is true that people are not perfectly aware of how the mechanism of their reasoning works and often they cannot pin down clearly the causes of their behaviour; yet they 'do have access to [their] goals, feelings and values', and through introspection they do have 'clues to the underlying process' of their own reasonings and calculations (Johnson-Laird and Shafir, eds, 1993). However, the process of thinking is a most complex one (Searle, 2001), in which both rationality and chance play a part. Neither is the inner working of his own mind fully grasped by the individual thinker, nor has cognitive science provided a full understanding of it so far.

Second, stressing the role that chance plays in human events does not destroy the aims of scientific rationality or the striving for an objective science. It is not the case that 'the disenchantment with reason makes for an idealization of the unknown [which] is potentially dangerous . . . [and] may also lead to a form of relativism' (Steuerman, 2000). On the contrary, there is another clear danger, namely that a disregard of the unknown, both in science and in everyday life, could produce a far less rational and objective science.

Now let me put forward my closing arguments. My main contention is that rationality as a principle upon which the agent's choice of action takes place can be only partial – it is limited. This is so because of the effects of chance. Furthermore, I do not think that, whenever a situation of having to choose an action presents itself, rationality will basically always be applied. In order to disprove such a claim, the procedure would appear to be the application of a battery of assumptions and auxiliary assumptions step by step, as follows:

A. Let us assume that chance affects 'the state of the world' in which a choice has to be made. Does this destroy the possibility of rational choice? No, since we assume also that the agent is aware of the actual 'state of the world', affected as it might be by chance, and he acts on that basis in a rational manner. In other words, the effects of chance simply change the baseline from which perfectly rational choice is possible.
B. Let us suppose that chance also affects 'the agent's own state', both physical and mental: that is, whether he is fatigued or in top shape, nervous or calm, in a good or bad mood, and so on. Here, too, we may assume that his 'existing state' will be the basis upon which, from his point of view and with his given faculties, he can choose rationally.
C. The next step is, however, one where we raise the possibility that an agent has a particular 'state of mind' in which he is unwilling or unable to evaluate the possible outcomes of his action and/or that he decides to act contrary to his own desire, for instance by using certain means which he knows will not achieve his desire. In

> parenthesis, let me state that I maintain that certain chance factors will have brought him to such thinking and decision. Can we now say that this particular 'personal state' provides the basis upon which he makes his choice and that from such a baseline his choice is still rational? Now, given his unwillingness or inability to calculate the results of his action, and even his willingness to act against his desire, can we say that the actual decision is 'the best' he could take and, therefore, his choice is rational? Of course, where there is no possibility of choosing freely and an action is forced upon an agent, it would be rational to act against his immediate desire, since the forced action would have the purpose of achieving a higher order desire, such as mere survival. To return to the free choice situation, however: where the agent decides to act in such a way that his desire will not be achieved, can we still see this as rational action from his particular baseline?

My answer is categorically no. For this amounts to *reductio ad absurdum*, making everything ultimately reducible to the fundamental doctrine, the global thesis of human rationality. This kind of argument is unacceptable, since it ultimately means that, no matter what, any choice and all choice is rational; it denies the possibility of irrational choice/action, so that rationality becomes meaningless. Thus we come full circle back to Searle's point that you cannot have rationality unless you also have the possibility of irrationality (see above, Searle, 2001). I hold, therefore, that the argument for universal rationality in any process of choice, as proposed in C, breaks down at some point. The breakdown occurs where the baseline becomes one where it is no longer possible to apply the very concept of rationality.

Finally, choice, whether rational or irrational, is linked at every stage and in all its dimensions to chance: whether in the acquiring and application of knowledge about existing alternatives, in evaluating their possible outcomes, in deciding on preferences, or in making the final choice. The question is how the effects of chance could be integrated into the study of the cognitive processes of motivation, goal-seeking, and intentionality leading to a choice of action. This seems to be an insuperable task, given the very nature of 'chance' as an unpredictable and uncontrollable element. It may well be that chance effects will remain the unknown factor in the equation of decision-making (like the residuals in correlational studies which can be defined as the differences between observed values and those predicted by a scientist's model; those differences representing the effect of variables that have not been examined or recognized by the scientist but which operate in addition to the ones included in the model). How to cope with chance within a research framework is a major challenge, providing yet another example of 'the limits of science' (Krausz, 2000).[2]

Concluding Remarks

I shall now draw my conclusions more concisely:

1 It is evident that both 'rationality', the mind's method of solving problems and dealing with choice situations, and 'chance' events, which are an integral part of reality, play vital roles in all that is connected with the human choice of action.
2 It is the intentionality of human action that strengthens our view of the necessity of rational thought in the choices we make; the corollary being that, if 'chance' is all pervasive, intentionality goes by the board. Yet the effects of 'chance' cannot be ignored. The concept of 'bounded rationality' must therefore be broadened by taking into account not only the limits of cognitive mental ability and the limits of knowledge, but also the effects of chance. The choice situation must not be seen as simply one where logical calculation alone determines the decision taken.
3 It is a truism that rationality is the element which links intentionality most clearly with goal-seeking. In other words, the individual agent is rational in so far as he is aware of the need to choose an action in order to attain his goal. However, he faces a more complex task where he has to decide what action to choose from a number of options available. Here he can rely on logical calculation, opt for

a guess, or even choose irrationally. In all these cases chance will have operated by influencing the states that determine his choice, both his physical and mental states and the state of the world he is facing. But, considering the likelihood that a rational way of thinking is adopted – which is one of the mind's primary capacities, especially where a choice/decision situation is faced – 'rationality' is still circumscribed by 'chance', since chance affects the very process that leads to the choice actually made. In sum, chance is the built-in factor in the open social world, with its constantly changing and interacting events and the resulting contingent causality; a world fuelled by the will of human beings who, by their choice of action, create the day-to-day reality.

Notes

1 My discussion of the possible definitions and meanings of the terms 'choice', 'rationality' and 'chance' is based largely on my article 'A Sociological Theory of Choice', *BDD – Journal of Torah and Scholarship*, Bar-Ilan University Press, No. 9, Summer 1999, pp. 24–8.

2 Despite the limitations mentioned, some comment on possible research on the influence of chance in decision-making will be in place. First, I wish to note that the large volume of research in psychology and economics on decision-making has been concerned with topics such as: why people violate rational decision principles; situational and personal constraints; limited cognitive capacity; decision strategies; regression analyses designed to determine which factors are most predictive; the possibility of teaching rational decision-making techniques; people's inherent information processing behaviour; the effect of market experience (Hardman and Harries, 2002). All this seems to have the aim to perfect or improve rational decision-making. Very little attention has been given to uncovering the pervasiveness of chance and its influence on decision-making. Comparative research could throw light on this topic by studying subsamples of populations who report on: (a) the use of mental rational computation in decision-making so as to arrive at the maximization of utility; (b) the reliance on chance impressions and guesses; (c) the use of rational calculation which is affected, however, by chance events. Such research could help us to evaluate the extent and

degree to which chance plays a part in the thinking process which leads to the choice of action.

Bibliography

Agor, Weston H., 1986. *The Logic of Intuitive Decision Making*, New York: Quorum Books, p. 6.

Arrow, Kenneth J., 1984. *Individual Choice under Certainty and Uncertainty*, Oxford: Basil Blackwell, p. 60.

Audi, Robert, 1993. *Action, Intention, and Reason*, Ithaca: Cornell University Press, pp. 4, 282.

Balzer, Noel, 1993. *The Human Being as a Logical Thinker*, Amsterdam–Atlanra, GA: Rodopi, pp. 61–4.

Baron, Jonathan, 1994. *Thinking and Deciding*, Cambridge: Cambridge University Press, pp. 3, 19–25, 36.

Bartlett, Frederic, 1958. *Thinking: An Experimental and Social Study*, London: Allen and Unwin, p. 20.

Beach, Lee Roy, 1997. *The Psychology of Decision Making*, Thousand Oaks, CA: Sage Publications, p. 4.

Bhaskar, Roy, 1989. *The Possibility of Naturalism*, New York: Harvester Wheatsheaf, pp. 25–6, 45, 80–1, 114.

Burks, Arthur W., 1977. *Chance, Cause, Reason*, Chicago: University of Chicago Press, Chapter 9.

Calhoun, Cheshire and Robert C. Solomon, eds, 1984. *What is an Emotion? Classic Readings in Philosophical Psychology*, New York/Oxford: Oxford University Press, pp. 306, 319–20.

Coren, S., and J. Miller, 1974. 'Size Contrast as a Function of Figural Similarity', *Perception and Psychophysics*, 16, pp. 355–7.

Crozier, Ray and Rob Ranyard, 1997. 'Cognitive Process Models and Explanations of Decision Making', in Rob Ranyard, Ray Crozier and Ola Svenson, *Decision Making*, London: Routledge, pp. 5, 6, 15.

Damasio, Antonio R., 1995. *Descartes' Error: Emotion, Reason and the Human Brain*, New York: Avon Books.

Damasio, Antonio R., 1999. *The Feeling of What Happens: Body and Emotion in the Making of Consciousness*, New York: Harcourt Brace.

Donagan, Alan, 1987. *Choice – The Essential Element in Human Action*, London: Routledge and Kegan Paul, pp. 9, 43.

Evans, Dylan & Pierre Cruse, 2004. *Emotion, Evolution and Rationality*, Oxford: Oxford University Press.

Evans, J. S. B. T., 1993. 'Bias and Rationality', in K. I. Manktelow and D. E. Over, eds, *Rationality: Psychological and Philosophical Perspectives*, London and New York: Routledge, pp. 6–30.

Evans, J. S. B. T., D. E. Over and K. I. Manktelow, 1994. 'Reasoning, Decision Making and Rationality', in P. N. Johnson-Laird and Eldar Shafir, eds, *Reasoning and Decision Making*, Cambridge, MA and Oxford, UK: Blackwell, pp. 166–87.

Fay, Brian, 1996. *Contemporary Philosophy of Social Science*, Oxford: Blackwell, p. 92.

Festinger, L., 1957. *A Theory of Cognitive Dissonance*, Evanston, IL: Row, Peterson.

Gellner, Ernest, 1992. *Reason and Culture*, Oxford: Blackwell, p. 49.

Gigerenzer, Gerd, Zeno Swijtink, Theodore Porter and Lorraine Daston, 1990. *The Empire of Chance*, Cambridge: Cambridge University Press, p. 220.

Ginet, Carl, 1997. 'Reasons Explanation of Action: An Incompatibilist Account', in Alfred R., Mele, ed., 1997, *The Philosophy of Action*, Oxford: Oxford University Press, pp. 106–13.

Glanz, Iosef, 1989. *Thinking*, Tel-Aviv: Gomeh Scientific Publications–Tcherikover Publishers, p. 43.

Gould, S. J., 1971. *The Rational Society*, London: The Athlone Press–University of London, pp. 1–3.

Hardman, David and Clare Harries, 2002. 'How Rational Are We?', *The Psychologist*, Vol. 15, Part 2 (February).

Hwang, Philip H., 1993. 'Aristotle and Peirce on Chance', in Edward C. Moore, ed., *Charles S. Peirce and the Philosophy of Science*, Tuscaloosa, Alabama: University of Alabama Press, p. 273.

Johnson, Donald M., 1972. *The Psychology of Thinking*, New York: Harper and Row, pp. 14–15.

Johnson-Laird, P. N., and Eldar Shafir, eds, 1993. *Reasoning and Decision Making*, Cambridge, MA and Oxford, UK: Blackwell, p. 3.

Kaufman, Bruce E., 1999. 'Expanding the Behavioral Foundations of Labor Economics', *Industrial and Labor Relations Review*, Vol. 52, No. 3 (April), pp. 378–9.

Khlentzos, Drew, 2008. 'Semantic Challenges to Realism', in N. Zalta, ed., *The Stanford Encyclopedia of Philosophy*, accessed at http://plato.stanford.edu/archives/fall2008/entries/realism-sem-challenge/ [original publication 2001, substantial revisions 2004].

Krausz, Ernest, 1999. 'A Sociological Theory of Choice', *BDD – Journal of Torah and Scholarship'*, Ramat-Gan: Bar-Ilan University Press, No. 9 (Summer), pp. 24–8.

Krausz, Ernest, 2000. *The Limits of Science*, New York: Peter Lang, pp. 47f., 53f., 61–6.

Levi, Isaac, 1986. *Hard Choices*, Cambidge: Cambridge University Press, p. 54.

Levi, Isaac, 1997. *The Covenant of Reason*, Cambridge: Cambridge University Press, Chapters 1, 2 and 4.

Loftus, E., *Memory*, 1980. New York: Ardsley House.

Malpas, J. E., ed., 1994. *The Philosophical Papers of Alan Donagan*, Vol. 2: *Action, Reason and Value*, Chicago: University of Chicago Press.

Manktelow, K. I., and D. E. Over, eds, 1993. *Rationality: Psychological and Philosophical Perspectives*, London and New York: Routledge.

Martin, Leonard L., 2001. 'Mood as Input: A Configural View of Mood Effects', in Martin, L. L. and Gerald L. Clore, eds, *Theories of Mood and Cognition*, Mahwah, New Jersey: Lawrence Erlbaum Associates, pp. 39, 63, 129, 135.

Martin, L. L., and Gerald L. Clore, eds, 2001. *Theories of Mood and Cognition*, Mahwah, New Jersey: Lawrence Erlbaum Associates.

Mellor, D. H., 1971. *The Matter of Chance*, Cambridge: Cambridge University Press, p. xi.

Milgram, S., 1974. *Obedience to Authority*, New York: Harper and Row.

Monod, Jacques, 1972. *Chance and Necessity*, London: Collins, p. 111.

Moser, Paul K., 1990. *Rationality in Action*, Cambridge: Cambridge University Press.

Myers .G., 1990. *Social Psychology* (2nd edition), New York: McGraw-Hill.

Nagel, Ernest, 1961. *The Structure of Science*, London: Routledge, pp. 324–35.

Newton, Roger G., 1997. *The Truth of Science*, Cambridge, MA: Harvard University Press.

Nussbaum, Martha, 2001. *Upheavals of Thought – The Intelligence of Emotions*, Cambridge: Cambridge University Press, pp. 1, 3, 90.

Plous, Scott, 1993.*The Psychology of Judgment and Decision Making*, New York: McGraw-Hill, Chapter 2.

Poincaré, Henri, 1905. *Science and Hypothesis*, New York: Walter Scott, p. 189.

Popper, Karl R., 1982. 'Quantum Theory and the Schism in Physics', Postscript to *The Logic of Scientific Discovery*, Vol. 3, London: Hutchinson.

Popper Karl, R., 1994. *The Myth of the Framework – In Defence of Science and Rationality*, London: Routledge, pp. 178f.

Popper, Karl R., 1995. *A World of Propensities*, Bristol: Thoemmes, p. 14.

Price, H. H., 1969. *Thinking and Experience*, London: Hutchinson, pp. 253–4.

Putnam, Hilary, 1983. *Realism and Reason – Philosophical Papers*, Vol. 3, Cambridge: Cambridge University Press, pp. 153f.

Radford, Jones and Andrew Burton, 1974. *Thinking: Its Nature and Development*, New York: John Wiley and Sons, pp. 59, 251.

Rubinstein, Ariel, 1998. *Modeling Bounded Rationality*, Cambridge, MA: MIT Press, pp. 2, 14, 16, 41, 52, 63, 88.

Rule, James B., 1997. *Theory and Progress in Social Science*, Cambridge: Cambridge University Press, pp. 83f.

Russell, Bertrand, 1940. *An Inquiry into Meaning and Truth*, London: Allen and Unwin, pp. 65, 167–9.

Searle, John R., 1983. *Intentionality*, Cambridge: Cambridge University Press.

Searle, John R., 1991. 'Response: Meaning, Intentionality and Speech Acts', in Ernest Lepore and Robert Van Gulick, eds, *John Searle and His Critics*, Oxford: Blackwell, p. 85.

Searle, John R., 2001. *Rationality in Action*, Cambridge, MA: MIT Press, pp. xiv, 5, 7, 23.

Sloman, Aaron, 1993. 'The Mind as a Control System', in Christopher Hookway and Donald Peterson, eds, *Philosophy and Cognitive Science*, Cambridge: Cambridge University Press, pp. 105–6.

Sfendoni-Mentzou, Demetra, 1993. 'The Role of Potentiality in Peirce's Tychism and in Contemporary Discussions in Quantum Mechanics and Microphysics', in Edward C. Moore, ed., *Charles S. Peirce and the Philosophy of Science*, Tuscaloosa, Alabama: University of Alabama Press, p. 253.

Shackle, G. L. S., 1958. *Time in Economics*, Amsterdam: North Holland, pp. 20–1.

Simon, Herbert A., 1982. *Models of Bounded Rationality*, Vol. 2, Cambridge, MA: MIT Press.

Smelser, Neil J., 1998. 'The Rational and the Ambivalent in the Social Sciences', *American Sociological Review*, Vol. 63, pp. 5, 6.

Solomon, Robert C., 2003. *Not Passion's Slave – Emotions and Choice*, New York/Oxford: Oxford University Press, p. 39.

Sripada, Chandra, and Stephen Stich, 2002. 'Evolution, Culture and the Irrationality of the Emotions', in Dylan Evans and Pierre Cruse, eds, *Emotion, Evolution and Rationality* (an interdisciplinary conference hosted by the Philosophy Department at King's College, London, 27–28 April 2002), accessed at http:// www.kcl.ac.uk/ip/nevensesardic, abstract,html.

Steuerman, Emilia, 2000. *The Bounds of Reason*, London and New York: Routledge, 2000, p. xvi.

Taton, R., 1957. *Reason and Chance in Scientific Discovery*, London: The Scientific Book Guild–Beaverbrook Newspapers/Hutchinson, pp. 79, 85, 86.

Tversky, Amos, and Daniel Kahneman, 1981. 'The Framing of Decisions and the Psychology of Choice', *Science*, Vol. 211, pp. 453–8.

Tversky, Amos and Daniel Kahneman, 1988. 'Rational Choice and the Framing of Decisions', in David E. Bell, Howard Raiffa and Amos Tversky, eds, *Decision Making*, Cambridge: Cambridge University Press, p. 172.

Velleman, J. David, 2000. *The Possibility of Practical Reason*, Oxford: Clarendon Press, Chapters 1, 6, 7.

Venn, John, 1962. *The Logic of Chance*, New York, NY: Chelsea Publishing Company, pp. 96, 246, 259.

Wakefield, Jerome, and Hubert Dreyfus, 1991. 'Intentionality and the Phenomenology of Action', in Ernest Lepore and Robert Van Gulick, eds, *John Searle and His Critics*, Oxford: Blackwell, pp. 261–3.

Weber, Max, 1964. *The Theory of Social and Economic Organization*, translated by A. M. Henderson and Talcott Parsons, New York: The Free Press/London: Collier–Macmillan, pp. 106f., 115–18.

Wicklund, R. A., and J. W. Brehm, 1976. *Perspectives on Cognitive Dissonance*, Hillsdale, NJ: Erlbaum.

Worth, L. T., and D. M. Mackie, 1987. 'Cognitive Mediation of Positive Mood in Persuasion', *Social Cognition*, 5, pp. 76–94.

6 The Human Condition – Does Reality Change?

In this concluding chapter I wish to pose the question of the sense in which reality can change and ask, particularly concerning the human world, whether reality has a flexibility that allows fundamental changes. There are two obstacles to our ability to give definitive answers. One is that change is inextricably bound up with time, a concept that we understand intuitively, but which, due to the irreversibility of time, prevents us from providing full and totally reliable evidence concerning past events and their effects. The second obstacle is that, specifically in the human context, the prediction of future events appears to be particularly unreliable. And yet much effort has gone into offering theoretical frameworks which propose to explain past evolutionary or developmental processes and to predict future ones.

To introduce the discussion around these questions, I wish to refer to John Searle's defence of *realism*, which he defines 'as the view that the world exists independently of our representations of it'. This view applies to 'external realism', which he also describes as the 'brute facts' of the world, or of the universe; a reality that exists irrespective of human existence and of statements or beliefs about it. In contrast to this kind of reality, there is a social reality which is totally dependent on human beings. Social reality is constituted by facts which exist only as a result of human belief and agreement, for instance property, citizenship, the exchange of goods, membership in families or clubs. These are institutionalized facts and they, too, exist objectively in that they are not dependent on any one human being's 'preferences, evaluations, or moral attitudes'. Concerning this dichotomy, one should also note 'the logical priority of brute facts over institutional facts' (Searle, 1995). Put more simply, human life *is* dependent on the brute physical facts.

Dividing reality in this way echoes similar distinctions made, as Karl Popper notes, at least since the times of the philosopher Protagoras in ancient Greece – namely 'the realization of the need to distinguish between two different elements in man's environment – his natural environment and his social environment'. Popper explains that in such distinctions it is important to stress the difference between *natural laws* such as the laws of gravity and thermodynamics, and *normative laws* such as the rules which forbid or demand certain modes of conduct. The former describe 'strict, unvarying regularity' and are beyond human control, whilst the latter depend on human actions and decisions (Popper, 1966).

Having noted the basic differences between the natural aspects and the human–social aspects of our world, what kind of changes do we observe in these realms? It will be sufficient to exemplify the constant changes we observe in the material aspects of our world. In the first place, the vastness we call 'universe' is constantly changing and expanding, according to the best knowledge of astrophysicists. In biological organisms there are constantly processes of change, which are seen in reproduction, growth, puberty and aging. In fact, birth and death mean that the actual human population living on this planet is not the same for any period of time in terms of its composition – it is constantly changing, and this change is never interrupted. The same applies to our flora and fauna. Of course, there is the constant change in weather, and also the serious change in the composition of the air due to pollution, which attests to the effect that human activity has on our physical surroundings. Yet there is also ample evidence in the natural world of uniformity and regularity: everything in the world consists of 'physical particles in fields of force' (Searle, 1995). But no less strong evidence of regularity, and even stability, concerning our small planet derives from facts such as that day follows night, that death follows life, that decay follows growth, and so on. And, in addition to these two complementary facets of regularity and change, there is also a third, no less important facet, which is the appearance of chaos and chance amidst regularity and change. This complex mixture makes our efforts to analyse reality, and to disentangle trends within it, an ever more daunting task.

I shall not pursue the topics of change and stability in the natural world. However, I do wish to make some comments on the changes that have occurred in the social and human world in the modern era, and to present some views on this sort of reality by highlighting certain events of the last century.

I have attempted in this book to see reality as it is, and not through the prism of some grand theory. More particularly, I have avoided the traps of historicism and historical materialism. The Hegelian and Marxian theories contend, as Popper puts it, 'that we can obtain any knowledge of social entities or essences only by applying the historical method, by studying social changes'. Thus '[c]hange, by revealing what is hidden in the underdeveloped essence, can only make apparent the essence, the potentialities, the seeds, which from the beginning have inhered in the changing object. This doctrine leads to the historicist idea of an historical fate or an inescapable essential destiny.' The fate of the human being, of the nation or of the state is sealed by its essence (Popper, 1966). Historical events, therefore, go along a natural path of development, over which there is no human control and which ends up, on the Hegelian and Marxian views, in some ideal condition for humanity. Thus, through historical stages or periods, primitive society develops into an advanced civilization, such as Marx's classless society.

The nineteenth century ushered in a tremendous popularity of modern historicist thinking among scholars dealing with human reality. This trend included Charles Darwin's evolutionary theory concerning living organisms, as well as the theories of August Comte and Karl Marx, who saw the various stages on the trajectory of history as leading to progress in a linear fashion, change having a definite direction and eventually culminating in an end-point (Van Den Berghe, 1961). For Comte, this process was to be the evolution of mankind to a state of scientific rationality, whilst for Marx the end-point would be the classless society: 'the class struggle would cease and the historical dialectic at this level will terminate' (Winter, 1964); Coser and Rosenberg, eds, 1976). Such utopian predictions, or rather prophecies, have not been borne out so far. In his book *The Poverty of Historicism*, Karl Popper has shown that, 'for strictly logical reasons, it is impossible for us to predict the future course of history'. 'The course of human history is strongly influenced by the growth of human knowledge.' However, 'we cannot predict, by rational or scientific methods, the future growth of our scientific knowledge'. Therefore Popper concluded that 'the historicist method is misconceived and collapses'. Furthermore, he argued that the human factor cannot be controlled by science, and if attempts were nevertheless made in this direction, then the objectivity of science would be destroyed and the freedom of individual diversity would be subjected to interference and oppressed. This would inevitably lead to an abandonment of rational thought and would be detrimental not only to the individual, but also to rationality and progress in human society (Popper, 2002).

Despite Popper's rather convincing arguments, evolutionary thinking has continued to influence many scholars throughout the twentieth century. They still believed that we have reached an epoch of advanced civilization, and not only in the fields of technology and science, but also in terms of culture and human society, which developed especially in the western world. I think that such views can no longer be upheld after the savagery that erupted in the twentieth century, particularly in Europe, which was regarded as the cradle of this advanced modern civilization. One only has to mention the two devastating world wars, the rise of relentlessly oppressive dictatorships, and – worst of all – the Holocaust.

How can we explain this development? Isaiah Berlin addresses this question, and especially the 'notion of an inevitable "march of history"'. While he expresses doubt as to the existence of laws of history, which would result in the realization of some preconceived idea about the future of civilization or in some particular social development in the future, he does accept the facts which have emanated from historical processes and have brought about advances in knowledge and great changes in our civilization. But we do not know what the true laws of such processes are. Furthermore, our belief that such advances in history will continue was 'rudely shaken by the evidence of the twentieth century'. Berlin totally rejects the Hegelian vision of human history 'as a logically necessary process of development passing through preordained stages to an ultimate rational goal where all human interests, capacities and values will coexist in frictionless harmony' (Hausheer, 1981). If anything, the events during the twentieth century have taught us that the techniques of modern civilization effectively helped those who wished to play on the irrational impulses of human beings, in their aim to defy the framework of civilized life, in order to achieve their evil intentions. Were the totalitarian regimes mere lapses in the inexorable laws which were supposed to lead to an advanced liberalizing civilization? Berlin disagrees with such a proposition. The events showed that a return to barbarism was possible. Men with sufficient energy and ruthlessness, who could collect sufficient power and who were ready to kill millions, could prevail. After all, Berlin argues, Hitler

> lost in the end [but] he lost by so narrow a margin that it does not need an eccentric imagination to conceive that he might have won, and that the consequences of his victory would have finally reduced to nonsense the doctrine according to which his rise and his victories were demonstrably impossible. (Berlin 1996)

In the end, modern civilization does not guarantee us protection 'against lapses into the past or violent lunges in unpredictable directions' (ibid.).

Let me adduce some more views concerning the disastrous twentieth century and its implications, especially the dilemmas it has produced for all of us who thought that we lived in a supposedly advanced civilization, but actually witnessed its unbelievable atrocities. First I quote S. N. Eisenstadt, who observes:

> Barbarism is not a vestige of premodern times, a survival of 'dark ages'. It is inherent in modernity: it epitomizes the dark side of modernity. Modernity bore within itself not only the various great visions of Emancipation or the great promises of continuous self-correction and expansion, but also very destructive possibilities – violence, aggression, war, and genocide. While barbarism as manifest in genocide and war can be found, at least potentially, in all human societies, it developed in modernity in some distinct –

> possibly its most gruesome ways. The Holocaust, which took place in the very center of modernity, has been the most extreme manifestation and symbol of the negative, destructive potentialities of modernity, of the barbarism lurking within the very core of modernity. (Eisenstadt, 2002)

Eisenstadt goes on to explain: in all human society the construction of social order is inevitably influenced by the consciousness of human finality, which gives rise to existential uncertainty or anxiety. Such a basic background produces 'tensions and confrontations between the more constructive and the more destructive potential inherent in the very creation of social order'. Collective identities are thus developed, establishing the borders and meanings of collectivities and the relations between them. In modern societies there was a distinct shift in conceiving of such collectivities – a shift towards minimizing the idea of the legitimacy of multiple individual and group goals and interests, totalitarian tendencies and the exclusiveness nurtured by the modern nation–state gaining the upper hand. This is how the dictatorial regimes succeeded in 'the ideologization of violence, terror and wars and the total ideological exclusivity and demonization of the excluded' (ibid.). Eisenstadt's analysis leads clearly to the conclusion that the barbarism of the twentieth century was a particular creation of modernity, which evolved with its detrimental tendencies and gruesome outcomes.

Not all would agree with the above analysis. Thus I. L. Horowitz does not go along with the conclusion that 'modern society itself is the "cause" of genocide', although he does raise the question why in the twentieth century societies had a penchant toward genocide and exhibited particular brutality (Horowitz, 1997). He rejects explanations, especially teleological ones, and the kind based on Rousseau's idea that advancing society pollutes and corrupts the human race. Neither modern society nor its highly developed technology are leading humanity solely to evil and destruction, if one stops to consider the positive effects of modernity in fields such as medicine and literacy.

Horowitz sees genocide as emerging where freedoms in the social order are obliterated by unbridled state power. 'Totalitarianism is the essence of the genocidal process.' The nagging question still remains whether one can see the Jewish Holocaust from a wider perspective of twentieth century genocides. Horowitz argues against such a view. However horrendous other genocides may have been, the Holocaust was unique in that its aim was the total annihilation of a people (Horowitz, 1997). No wonder that Auschwitz has become an indelible stain on Germany's history (*The Jerusalem Post*, 20 February 2008) and the unwanted mark of modern European civilization. Jean-François Lyotard maintains that, as a consequence of the twentieth-century disasters, members of contemporary societies

> no longer find it possible to believe in their history as an epic of progress or emancipation. The promises of prosperity, freedom and justice associated with the Enlightenment project of scientific control over nature and a rational organization of society have signally failed to be realized, and we have no reason to suppose that they will be in the future. (Dews, 1992)

Modern society, with its hopes, 'has been liquidated' (ibid.).

What followed next? Is it true that, after 'the end of ideology', we are witnessing a new era, which signals major changes and is subsumed by philosophers and social scientists under labels such as 'the post-war world', 'the post-industrial society' or 'the post-modern condition'? (Bell, 1999; Lyotard, 1984). These changes and events, characterizing the new reality, are still being observed, interpreted and discussed. One thing is clear, though: by concentrating on transformations primarily in the technological and economic fields and on changes in our physical conditions, the new nomenclatures have brought about a distancing from the preceding, self-induced, calamities of mankind. Yet elements, and even repetitions, of such calamities as genocides and dictatorships seem to continue; and the increase in nuclear arming and terrorism has not been checked.

On the basis of the above views, I wish to stress the fact that modern reality, which has led to many positive achievements, especially in economy, science, medicine, technology and the arts, has also encouraged various evils, immorality and other negative results. And even some of the positive elements have boomeranged, so that, with the new scientific and technological knowledge, the possibilities of destructiveness have risen to mind-boggling catastrophic levels. Does this signal that contemporary reality has changed radically for the worse? And what is the angle from which we should view the prospects of humanity: a pessimistic outlook concerning human nature and social relations, or a basically hopeful, liberal–humanist conception of the individual and of his future in human society? (Hausheer, 1981). The answer cannot be unequivocal, since there is a basic difficulty in determining which age in human history exhibited in the past, or exhibits now, a more unbearable or dangerous condition for human existence. And what is unbearable or dangerous can be defined in various ways. Similarly, there is no objective assessment by which one can determine what was, or is, the direction of social change. The concepts involved – such as stability, progress, stagnation or retrogression – can be variously assessed and applied to different aspects and situations in human societies. Above all, the elements of uncertainty have always been, and still are, an integral part both of the natural realm and of the human world. My dwelling on these uncertainties, on the limits of knowledge, and on living with the unknown may create the impression that my assessment of reality is a sceptical one, which belittles the possibilities of a change towards a brighter future for mankind.

Of course, different approaches and assessments could lead to more optimistic views. Jeffrey Alexander's study of the civil sphere is perhaps the most convincing. His hope is grounded in the basic element of human solidarity. He recognizes, however, the less than solidary relations often exhibited in society, especially by the nation–state with its exclusionary tendencies, shown by the prevalence of anti-Semitism in Europe as well as in America. By the middle of the twentieth century, in a Europe dominated by Nazi Germany, anti-Semitism led to the most heinous of crimes, hitherto unequalled in human history: the Holocaust. Alexander is consequently aware of the possibility that the civil sphere can become fragmented and weakened, and can even totally collapse. Yet he sees the struggles to incorporate out-groups such as the Jews and other ethnic communities in America as an encouraging sign: a 'repair' and a strengthening of the democratic structure of civil society in real terms, and not just a utopian aspiration. However, Alexander also takes into account the element of uncertainty as to the direction which the civil sphere will take, considering that 'it will still be restless, and its dynamism will be dangerous and contradictory'. Nevertheless, in his view, the universal humane yearning for liberty and solidarity 'inspires hope for a democratic life' (Alexander, 2006). Jürgen Habermas' theory of communicative action (see Chapter 4) similarly tries to provide some hope for the future of society by advocating a rational approach in human relations, one designed to foster consensus rather than conflict in human affairs. Another clear example of hopefulness is expressed by Ernst Cassirer, who believes in 'man's progressive self-liberation' and in his 'power to build up a world of his own, an "ideal" world' (Cassirer, 1969). In contrast, my explorations of reality have led me to an assessment of the future which is not imbued with such hopes, and which offers the reader a rather despondent but realistic conclusion: reality is constantly changing, yet in its essence it remains the same.

Bibliography

Alexander, Jeffrey, C., 2006. *The Civil Sphere*, Oxford: Oxford University Press, pp. 142, 171–9, 293, 417, 468, 500–50, 553.

Berlin, Isaiah, 1996. *The Sense of Reality*, London: Chato and Windus/Pimlico, pp. 10–11.

Bell, Daniel, 1999. *The Coming of Post-Industrial Society*, New York: Basic Books.

Cassirer, Ernst, 1969. *An Essay on Man*, New Haven: Yale University Press, p. 228.

Coser, Lewis A., and Bernard Rosenberg, eds, 1976. *Sociological Theory*, New York: Macmillan and London: Collier–Macmillan, pp. 592–3.

Dews, Peter, 1992. 'Introduction', in Jürgen Habermas, *Autonomy and Solidarity. Interviews with Jürgen Habermas*, edited, introduced and with translation revised by Peter Dews, London–New York: Verso, pp. 5, 25.

Eisenstadt, S. N., 2002. 'Barbarism and Modernity: The Destructive Components of Modernity – The Perennial Challenge', in Ernest Krausz and Gitta Tulea, eds, *Starting the Twenty-First Century*, New Brunswick and London: Transaction Publishers, pp. 25–35.

Hausheer, Roger, 1981. 'Introduction', in Isaiah Berlin, *Against the Current*, Oxford: Oxford University Press.

Horowitz, Irving Louis, 1997. *Taking Lives – Genocide and State Power*,

New Brunswick and London: Transaction Publishers, pp. 19, 48, 250, 256.

LaPiere, Richard T., 1965. *Social Change*, New York: McGraw-Hill.

Lyotard, Jean-François, 1984. *The Postmodern Condition: A Report on Knowledge*, Minneapolis: University of Minnesota Press.

Popper, Karl R., 1966. *The Open Society and Its Enemies*, Vol. 2, Princeton, New Jersey: Princeton University Press, pp. 57–8.

Popper, Karl R., 2002. *The Poverty of Historicism*, London: Routledge Classics, pp. xi–xiii, 147.

Searle, John R., 1995. *The Construction of Social Reality*, New York: The Free Press, pp. 1–11, 55–6, 149–54.

Strawson, P. F., 1985. *Skepticism and Naturalism: Some Varieties*, New York: Columbia University Press.

Van Den Berghe, Pierre L., 1961. 'Dialectic and Functionalism – Toward a Theoretical Synthesis', in Walter L. Wallace, ed., *Sociological Theory*, London: Heinemann, p. 210.

Winter, A. B., 1964. 'Class Struggle', in Julius Gould and William L. Kolb, eds, *A Dictionary of the Social Sciences*, New York: Free Press of Glencoe, p. 96.

Name Index

Name Index

Name Index

Subject Index

Subject Index